Crystal Growth and Chirality - Technologies and Applications

Edited by Riadh Marzouki and Takashiro Akitsu

Published in London, United Kingdom

Crystal Growth and Chirality - Technologies and Applications
http://dx.doi.org/10.5772/intechopen.97896
Edited by Riadh Marzouki and Takashiro Akitsu

Contributors
Senthil Kumar Chandran, Chinnakannu Elavarasi, John James Gnanapragasam, Srinivasan Manikam, Hendrik J. R. Lemmer, Wilna Liebenberg, Allen Moses Samuel Elizabeth, Riadh Marzouki, Basma Marzougui, Amira Marzouki, Youssef Ben Smida, Claudia Garnero, Agustina Bongioanni, Maria Soledad Bueno, Belén Alejandra Mezzano, Marcela Raquel Longhi, Gayani Pathiraja, Sherine Obare, Hemali Rathnayake, Lamia Khedhiri, Akira Kouchi, Takashi Shimonishi, Tomoya Yamazaki, Masashi Tsuge, Naoki Nakatani, Kenji Furuya, Hiromasa Niinomi, Yasuhiro Oba, Tetsuya Hama, Hiroyasu Katsuno, Naoki Watanabe, Yuki Kimura, Toru Asahi, Kenta Nakagawa, Masahito Tanaka, Yukana Terasawa, Kazuhiko Ishikawa, Akifumi Takanabe, Hideko Koshima, Bart Kahr

First published in London, United Kingdom, 2023 by IntechOpen
IntechOpen is the global imprint of INTECHOPEN LIMITED, registered in England and Wales, registration number: 11086078, 5 Princes Gate Court, London, SW7 2QJ, United Kingdom

British Library Cataloguing-in-Publication Data
A catalogue record for this book is available from the British Library

Additional hard and PDF copies can be obtained from orders@intechopen.com

Crystal Growth and Chirality - Technologies and Applications
Edited by Riadh Marzouki and Takashiro Akitsu
p. cm.
Print ISBN 978-1-80355-057-2
Online ISBN 978-1-80355-058-9
eBook (PDF) ISBN 978-1-80355-059-6

Meet the editors

Riadh Marzouki is an associate professor in the Department of Chemistry, Faculty of Sciences at King Khalid University, Saudi Arabia, an international collaborator in the Department of Chemistry, Faculty of Sciences, University of Sfax, Tunisia, and a researcher in the Laboratory of Materials and Crystallography of the University of Tunis El-Manar, Tunisia. His research interests include the synthesis, crystal structure, electrical properties, and biological activities of new crystalline materials. He is the author and co-author of more than 80 scientific papers, four book chapters, and three edited books.

Takashiro Akitsu, Ph.D., is a professor in the Department of Chemistry, Faculty of Science Division II, Tokyo University of Science, Japan. He studied at the Institute for Protein Research, Osaka University (metalloproteins), Keio University, Japan (photo and magnetic functional organic/inorganic hybrid compounds), and Stanford University, USA (physical and bioinorganic chemistry). Dr. Akitsu obtained his Ph.D. in physical and inorganic chemistry from Osaka University, Japan, in 2000. He has published 220 articles and book chapters. He has also served as an editorial board member and peer reviewer for many journals and has been involved in the organizing committees for several international conferences.

Contents

Preface **IX**

Section 1
Synthesis and Crystal Structure Determination: Examples **1**

Chapter 1 **3**
Crystal Structure of $[C_{17}H_{22}N_2]_3[P_6O_{18}][H_2O]_8$
by Lamia Khedhiri

Chapter 2 **15**
The Cuprate Ln_2CuO_4 (Ln: Rare Earth): Synthesis, Crystallography, and Applications
by Basma Marzougui, Amira Marzouki, Youssef Ben Smida and Riadh Marzouki

Chapter 3 **35**
Pharmaceutical Crystals: Development, Optimization, Characterization and Biopharmaceutical Aspects
by Agustina Bongioanni, Maria Soledad Bueno, Belén Alejandra Mezzano, Marcela Raquel Longhi and Claudia Garnero

Section 2
Crystallization and Crystal Growth **55**

Chapter 4 **57**
Oriented Attachment Crystal Growth Dynamics of Anisotropic One-dimensional Metal/Metal Oxide Nanostructures: Mechanism, Evidence, and Challenges
by Gayani Pathiraja, Sherine Obare and Hemali Rathnayake

Chapter 5 **83**
Recent Advances in Infrared Nonlinear Optical Crystal
by Senthil Kumar Chandran, Chinnakannu Elavarasi, Srinivasan Manikam and John James Gnanapragasam

Chapter 6 **105**
Crystallization: Its Mechanisms and Pharmaceutical Applications
by Hendrik J.R. Lemmer and Wilna Liebenberg

Section 3
Chirality **123**

Chapter 7 **125**
Chiral Ice Crystals in Space
by Akira Kouchi,Takashi Shimonishi,Tomoya Yamazaki, Masashi Tsuge, Naoki Nakatani, Kenji Furuya, Hiromasa Niinomi, Yasuhiro Oba, Tetsuya Hama, Hiroyasu Katsuno, Naoki Watanabe and Yuki Kimura

Chapter 8 **145**
Chiroptical Studies on Anisotropic Condensed Matter: Principle and Recent Applications of the Generalized-High Accuracy Universal Polarimeter
by Toru Asahi, Masahito Tanaka, Kenta Nakagawa, Yukana Terasawa, Kazuhiko Ishikawa, Akifumi Takanabe, Hideko Koshima and Bart Kahr

Chapter 9 **167**
Third-Order Nonlinearity Measurement Techniques
by Allen Moses Samuel Elizabeth

Preface

Crystallography is an important field of science that deals with the arrangement and bonding of atoms in crystalline solids and the geometric structure of crystal lattices. It is useful in phase identification, which is important for using a material, for example in manufacturing, as the composition, structure, and proportions of the phases present in the material will influence its properties. Crystallography is also essential in determining the 3D structures of biological macromolecules such as proteins and nucleic acids, which are important for drug discovery and development. Additionally, crystallography plays a significant role in modern living, particularly in electronics, where the properties of crystals are fundamental to the technology used. In this book, we deal with some crystalline materials synthesized by different methods, their crystal growth in relation to their physical properties or biological activities, and their applications, in particular in biology, medicine, and semiconductors.

Dr. Riadh Marzouki
Associate Professor,
Chemistry Department,
College of Science,
King Khalid University,
Abha, Saudi Arabia

Researcher,
Laboratory of Materials,
Crystal Chemistry, and Applied Thermodynamics,
Faculty of Sciences of Tunis,
University of Tunis El Manar,
Tunis, Tunisia

International Collaborator,
Chemistry Department,
Faculty of Science of Sfax,
Sfax, Tunisia

Dr. Takashiro Akitsu
Professor,
Department of Chemistry,
Tokyo University of Science,
Tokyo, Japan

Section 1

Synthesis and Crystal Structure Determination: Examples

Chapter 1

Crystal Structure of $[C_{17}H_{22}N_2]_3[P_6O_{18}][H_2O]_8$

Lamia Khedhiri

Abstract

$[C_{17}H_{22}N_2]_3[P_6O_{18}][H_2O]_8$., a new organic cyclohexaphosphate, was synthesized as single crystals and grown by solution growth method. The crystal structure of the grown product was determined by single crystal X-ray diffraction analysis. The title material crystallized in the monoclinic system of the C2/c space group. The $P_6O_{18}{}^{6-}$ ring anions and some water molecules form layers spreading around (b, c) planes via O-H…O hydrogen bonds. Between these inorganic layers are anchored organic cations, which establish H-bonds to interconnect the different adjacent layers and so contribute to the cohesion of the three dimensional lattice. The organic and inorganic rings adopt a chair conformation with different geometrical characteristics due to their different size and flexibility. The title compound was further characterized by FT-IR and NMR spectroscopy.

Keywords: cyclohexaphosphate, crystal structure, inorganic layers, three dimensional lattice

1. Introduction

Since the preparation and identification of $Li_6P_6O_{18}.6H_2O$ [1], this salt was used as starting reagent to prepare other cyclohexaphosphates. These latters were of great interest in academic and industrial areas over recent years owing to their diversities [2, 3]. Among these compounds, organic cyclohexaphosphates are particularly interesting. Both H-bonds and the organic units are responsible in the organization of such materials as to build inorganic lattices with different features: ribbons [4], two-dimensional lattices [5–7], and three-dimensional lattices [8, 9]. To study the effect on the chemical and structural geometries, we report and discuss in the present work the results of a structural investigation concerning a new organic-cation cyclohexaphosphate $[C_{17}H_{22}N_2]_3[P_6O_{18}][H_2O]_8$. This latter was also characterized by IR and NMR spectroscopy.

The selected organic molecule, 1-(diphenylmethyl)piperazine, is a biogenic diamine, which plays an important role as a deprotonated cation in biological systems [10, 11].

2. Chemical preparation

Based in previous work, the LiH_2PO_4 [12] and $Li_6P_6O_{18}.6H_2O$ [13] salts were synthesized first. Then, the title compound was prepared by adding cyclohexaphosphoric acid dropwise, produced from $Li_6P_6O_{18}.6H_2O$ through a cation-exchange resin (Amberlite IR 120) into an ethalonic solution of 1-(diphenylmethyl)piperazine. The obtained solution was stirred for few minutes and allowed to a slow solvent evaporation at room temperature until stable crystals of $[C_{17}H_{22}N_2]_3[P_6O_{18}][H_2O]_8$ with suitable dimensions were obtained.

3. Structure description

Crystal data and details of structure refinement are summarized in **Table 1**. Molecular graphics were performed employing Diamond [14]. The chemical composition of $[C_{17}H_{22}N_2]_3[P_6O_{18}][H_2O]_8$ includes three entities, one phosphoric ring, eight water molecules and three crystallographically distinct organic cations. An ORTEP view of the geometrical configuration of these entities is depicted in **Figure 1**, while

Empirical formula	$C_{51}H_{82}N_6O_{26}P_6$
Formula weight [g mol^{-1}]	1381.04
Crystal color, habit	colorless, rod
Crystal temperature [K]	295
Crystal size [mm^3]	0.15 × 0.23 × 0.49
Radiation, wavelength [Å]	Mo $K\alpha$, λ = 0.71073
Crystal system	Monoclinic
Space group	C2/c
Unit-cell dimensions:	
a, *b*, *c* [Å]	36.0432(10), 12.9785(4), 34.1510(9)
β [°]	118.7840(8)
Volume [Å^3]	14001.5(7)
Z	8
Density calc. [g cm^{-3}]	1.310
Reflections for cell determination	26,969
θ range for cell determination [°]	2 to 27
Absorption coefficient μ [mm^{-1}]	0.232
F(000)	5824
θ-Range for data collection [°]	1.289 to 26.999
Limiting indices	$-45 \leq h \leq 46$, $-16 \leq k \leq 15$, $-43 \leq l \leq 43$
Reflections collected/unique	26,969, 10,651 (*R*int = 0.033)
Refinement method	Full-matrix least-squares on F^2
Data, restrains, parameters (I > 2 σ)	15,155, 24, 851
Goodness-of-fit on F^2	1.094
R indices (all data, on F^2)	R = 0.0663, Rw = 0.2133

Table 1.
The crystal data and experimental parameters used for the intensity data collection. Procedure and final results of the structure determination.

DOI: http://dx.doi.org/10.5772/intechopen.108160

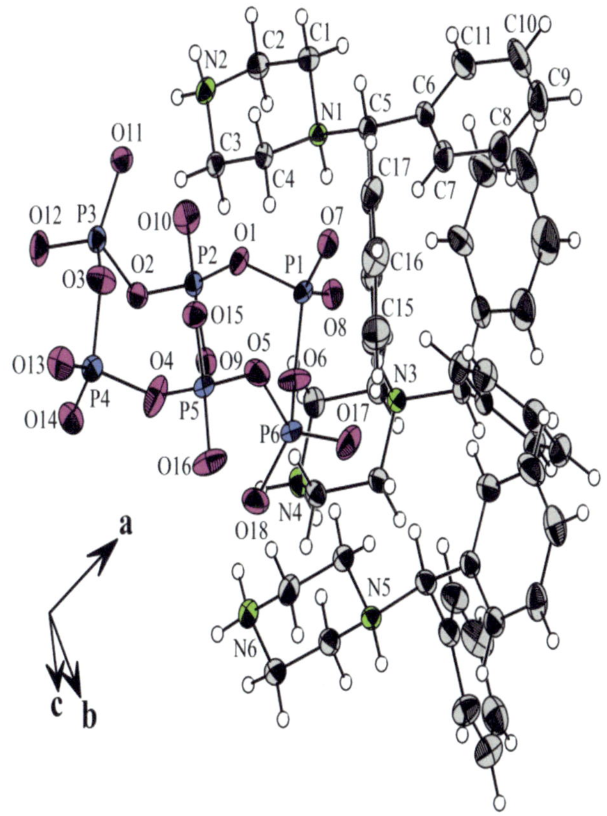

Figure 1.
ORTEP Plot of phosphoric ring and independent organic cations of $[C_{17}H_{22}N_2]_3[P_6O_{18}][H_2O]_8$ *with numbering scheme. Thermal ellipsoids are drawn at 40% of probability. The water molecules are omitted for figure clarity.*

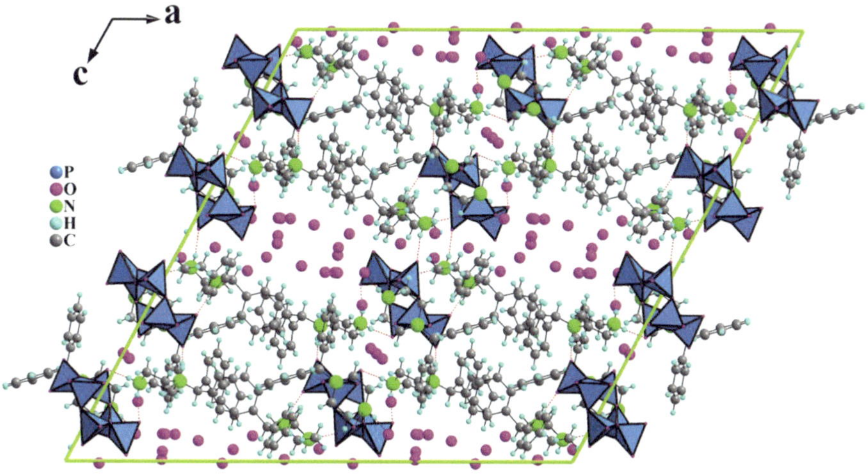

Figure 2.
Projection of the structure of $[C_{17}H_{22}N_2]_3[P_6O_{18}][H_2O]_8$*, along the b axis, The phosphoric anions are given in tetrahedral representation. Hydrogen bonds are shown as dashed lines.*

Figure 2 exhibits the complete atomic arrangement. The packing of the title compound consists of hybrid layers where the organic and inorganic species are alternated. Theses layers, extended perpendicularly to the b-axis, are also connected by H-bonds in the two other directions assuring the cohesion of the lattice.

A chair conformation for the P_6O_{18} ring anion was adopted (**Figure 3-a**). This phosphoric ring includes six independent PO_4 tetrahedra. The values of the P−O−P, O−P−O angles and the P−O and O−O distances, are listed in **Table 2**. The P−O bond lengths vary between 1.457 and 1.612 Å with an average value of 1.534 Å. The variation of the environment around the oxygen atoms can explains this divergence. Despite this diversity in P−O distances, each tetrahedron in the P_6O_{18} anion can be represented by typical O atoms arrangement with phosphorus atom moved of 0.117, 0.134, 0.138, 0.134, 0.147, 0.147 Å from the centre (**Table 3**, **Figure 3-a**). In addition, the distortion indices (DI) [15]: DI(PO) = 0.039, DI(OPO) = 0.038 and DI(OO) = 0.014 show an above distortion of the O-O bond lengths compared to P-O and O-P-O ones as illustrated in **Table 3**.

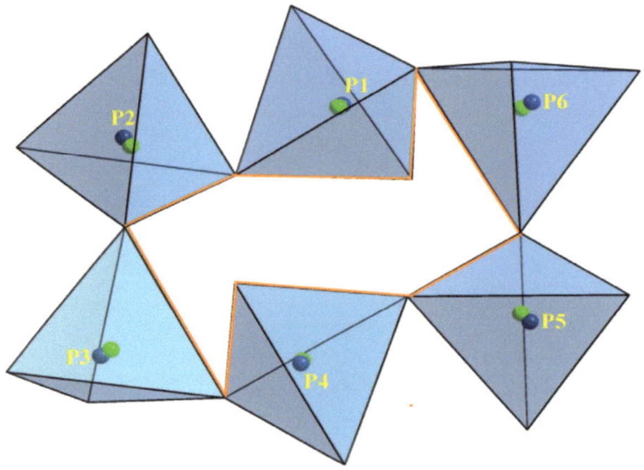

(a)

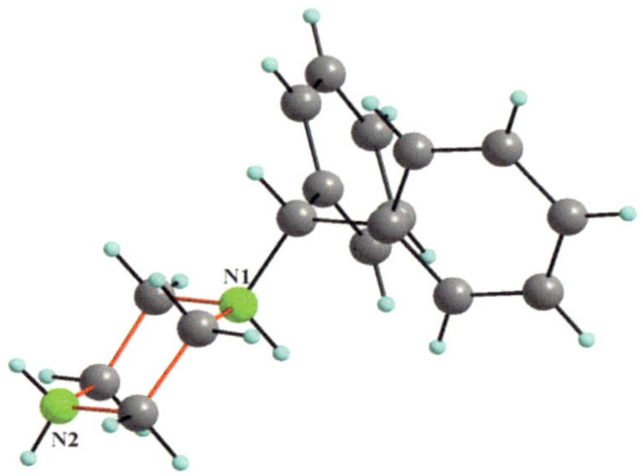

(b)

Figure 3.
Chair conformation of the inorganic and organic rings.

$P(1)O_4$ tetrahedron				
P1	O1	O6	O7	O8
O1	1.596(2)	103.49(16)	106.27(13)	108.74(15)
O6	2.491(0)	1.576(3)	111.97(15)	106.60(15)
O7	2.458(0)	2.530(0)	1.475(2)	118.70(15)
O8	2.501(0)	2.451(0)	2.541(0)	1.479(2)
$P(2)O_4$ tetrahedron				
P2	O1	O2	O9	O10
O1	1.599(2)	100.68(15)	111.09(14)	108.85(17)
O2	2.454(1)	1.589(2)	105.57(14)	110.48(17)
O9	2.541(1)	2.445(1)	1.481(2)	118.64(18)
O10	2.495(1)	2.511(1)	2.535(1)	1.467(3)
$P(3)O_4$ tetrahedron				
P3	O2	O3	O11	O12
O2	1.604(2)	100.96(16)	110.90(15)	106.44(16)
O3	2.470(0)	1.597(3)	107.62(15)	109.90(18)
O11	2.539(0)	2.481(0)	1.476(3)	119.46(17)
O12	2.465(0)	2.513(0)	2.546(0)	1.471(3)
$P(4)O_4$ tetrahedron				
P4	O3	O4	O13	O14
O3	1.601(3)	102.1(2)	106.09(17)	110.70(16)
O4	2.472(0)	1.579(3)	111.23(19)	104.78(17)
O13	2.455(0)	2.517(0)	1.470(3)	120.52(17)
O14	2.539(0)	2.428(0)	2.565(0)	1.484(3)
$P(5)O_4$ tetrahedron				
P5	O4	O5	O15	O16
O4	1.593(3)	98.30(18)	111.44(17)	107.9(2)
O5	2.404(0)	1.586(3)	107.85(14)	110.12(18)
O15	2.536(0)	2.508(0)	1.475(2)	119.24(17)
O16	2.479(0)	2.542(0)	2.474(0)	1.473(3)
$P(6)O_4$ tetrahedron				
P6	O5	O6	O17	O18
O5	1.612(3)	110.93(16)	108.42(18)	110.93(16)
O6	2.464(0)	1.596(3)	109.45(19)	105.94(18)
O17	2.491(0)	2.494(0)	1.457(3)	119.93(19)
O18	2.551(0)	2.460(0)	2.546(0)	1.483(3)

P1-P2	2.893(2)	P1-O1-P2	129.73(3	P1-P2-P3	114.21(2)
P2-P3	2.956(2)	P2-O2-P3	135.52(3)	P2-P3-P4	103.08(2)
P3-P4	2.931(2)	P3-O3-P4	132.9(3)	P3-P4-P5	110.47(2)
P4-P5	2.954(2)	P4-O4-P5	137.24(3)	P4-P5-P6	110.01(2)
P5-P6	2.909(2)	P5-O5-P6	130.97(3)	P5-P6-P1	104.39(2)
P6-P1	2.982(2)	P6-O6-P1	140.09(3)	P6-P1-P2	103.58(2)

Table 2.
Main interatomic distances (Å) and bond angles (°) in $[C_{17}H_{22}N_2]_3[P_6O_{18}][H_2O]_8$.

Tetrahedron	P-O$_m$	ID (P-O)	(O-P-O)$_m$	ID (OPO)	O-O$_m$	ID (O-O)	δ
$P_1(O_4)$	1.532	0.036	109.295	0.037	2.495	0.011	0.117
$P_2(O_4)$	1.534	0.039	109.251	0.038	2.497	0.013	0.134
$P_3(O_4)$	1.537	0.041	109.213	0.039	2.502	0.012	0.138
$P_4(O_4)$	1.534	0.037	109.236	0.045	2.496	0.017	0.134
$P_5(O_4)$	1.532	0.038	109.141	0.041	2.491	0.015	0.147
$P_6(O_4)$	1.537	0.044	110.93	0.027	2.501	0.013	0.147

Table 3.
Interatomic PO and OO distances (Å), OPO angles (°), tetrahedral distortion indexes ID(PO), ID(OPO) and ID(OO) of the cyclohexaphosphate in $[C_{17}H_{22}N_2]_3[P_6O_{18}][H_2O]_8$. The last column corresponds to the shift parameter.

D—H···A	D—H	H···A	D···A	D—H···A
N(1)-H(1)...O(7)	0.98	1.70	2.6681	168
N(2)-H(2A)...O(11)	0.89	1.90	2.7200	152
N(2)-H(2B)...O(15)	0.89	1.93	2.7408	151
N(3)-H(3)...O(8)	0.98	1.68	2.6354	164
N(4)-H(4C)...O(2)	0.89	2.59	2.9864	108
N(4)-H(4C)...O(9)	0.89	1.84	2.7231	170
N(4)-H(4D)...O(14)	0.89	1.89	2.7444	161
N(5)-H(5)...O(10)	0.98	1.69	2.6560	166
N(6)-H(6A)...O(16)	0.89	1.98	2.7625	147
N(6)-H(6B)...O(1 W)	0.89	1.85	2.6998	159
C(1)-H(1B)...O(13)	0.97	2.43	3.3031	149
C(3)-H(3A)...O(1)	0.97	2.54	3.4332	153
C(4)-H(4A)...O(15)	0.97	2.58	3.2464	126
C(5)-H(5A)...O(13)	0.98	2.44	3.3306	150
C(7)-H(7)...O(7)	0.93	2.38	3.2627	158
C(13)-H(13)...O(7)	0.93	2.59	3.4217	150
C(19)-H(19A)...O(9)	0.97	2.60	3.3733	137
C(20)-H(20B) ..O(6)	0.97	2.46	3.3110	147
C(22)-H(22) ..O(10 W)	0.98	2.52	3.4419	156
C(30)-H(30) ..O(8)	0.93	2.53	3.3191	142
C(35)-H(35A) ..O(18)	0.97	2.59	3.4981	156
C(37)-H(37A) ..O(11)	0.97	2.50	3.3556	148
C(37)-H(37B) ..O(8 W)	0.97	2.56	3.1961	123
C(39)-H(39) ..O(17)	0.98	2.40	3.3125	154

Table 4.
Hydrogen-bond geometry (Å, °) in $[C_{17}H_{22}N_2]_3[P_6O_{18}][H_2O]_8$.

The P-O-P angle values match well with those noticed in others cyclohexaphoshates [16, 17]. But the P1-O6-P6 angle of 140.09°, diverge from the value generally observed in such anions. This angle induces a longer P–P distance (2.982°).

DOI: http://dx.doi.org/10.5772/intechopen.108160

It should be signaled that these values display the greatest discrepancy measured until now. The P–P–P angles ranging from 103.08 to 114.21° which averages are 107.62°, show large deviations from the ideal value (120°). In spite of this variance, the distortion is more important if compared with that observed in $[C_9H_{14}N]_4[H_3O]_2[P_6O_{18}]$ [18] with the same space $C_2\backslash c$ group in which the average of the P–P–P angle is 101.0°.

In the crystal structure there are three independent 1-(diphenylmethyl)piperazinium cations that are associated with phosphoric entities through electrostatic interactions and hydrogen bonds involving hydrogen atoms of NH and NH_2. Each six-membered piperazinedium ring adopts a chair conformation (**Figure 3-b**). In all hydrogen bonds, the nitrogen atoms are donors, whereas the oxygen from the $P_6O_{18}{}^{6-}$ acts as acceptor atoms (**Table 4**), with N...O separations ranging from 2.6354 to 2.9864 Å. The (N-C, C-C) bond lengths and bond angles (N-C-C, C-C-C) ranging from 1.346(9) to 1.528(4) Å and from 108.4(3) to 124.6(4) ° (**Table 5**) are comparable with those observed in other organic phosphates [19].

$[C_{17}H_{22}N_2(1)]^+$ group			
N1 - C4	1.499(4)	C4 - N1 - C1	108.7(3)
N1 - C1	1.506(4)	C4 - N1 - C5	110.6(2)
N1 - C5	1.528(4)	C1 - N1 - C5	111.4(2)
N2 - C2	1.478(4)	C2 - N2 - C3	110.7(3)
N2 - C3	1.483(4)	N1 - C1 - C2	110.9(3)
C1 - C2	1.514(5)	N2 - C2 - C1	111.4(3)
C3 - C4	1.495(4)	N2 - C3 - C4	110.6(3)
C5 - C6	1.510(5)	C3 - C4 - N1	110.3(3)
C5 - C12	1.521(5)	C6 - C5 - C12	111.1(3)
C6 - C11	1.388(5)	C6 - C5 - N1	113.1(3)
C6 - C7	1.389(5)	C12 - C5 - N1	110.0(3)
C7 - C8	1.395(5)	C11 - C6 - C7	118.5(4)
C8 - C9	1.378(8)	C11 - C6 - C5	117.5(4)
C9 - C10	1.368(8)	C7 - C6 - C5	123.4(3)
C10 - C11	1.385(7)	C6 - C7 - C8	120.7(4)
C12 - C17	1.382(5)	C9 - C8 - C7	120.1(5)
C12 - C13	1.399(5)	C10 - C9 - C8	119.4(4)
C13 - C14	1.374(6)	C9 - C10 - C11	121.2(5)
C14 - C15	1.382(7)	C10 - C11 - C6	120.2(5)
C15 - C16	1.399(7)	C17 - C12 - C13	119.1(3)
C16 - C17	1.375(6)	C17 - C12 - C5	119.5(3)
	C9 - C10 - C11	121.2(5)	
	C10 - C11 - C6	120.2(5)	
	C13 - C12 - C5	121.4(3)	
	C14 - C13 - C12	120.4(4)	
	C13 - C14 - C15	120.3(4)	

	C14 - C15 - C16	119.6(4)	
	C17 - C16 - C15	119.9(4)	
	C16 - C17 - C12	120.7(4)	
$[C_{17}H_{22}N_2(2)]^+$ group			
N3 - C18	1.496(4)	C18 - N3 - C21	109.3(3)
N3 - C21	1.496(4)	C18 - N3 - C22	110.3(3)
N3 - C22	1.521(4)	C21 - N3 - C22	110.8(3)
N4 - C20	1.473(5)	C20 - N4 - C19	110.6(3)
N4 - C19	1.482(5)	N3 - C18 - C19	111.1(3)
C18 - C19	1.509(5)	N4 - C19 - C18	110.3(3)
C20 - C21	1.514(5)	N4 - C20 - C21	110.0(3)
C22 - C29	1.511(5)	N3 - C21 - C20	111.9(3)
C22 - C23	1.523(5)	C29 - C22 - N3	112.3(3)
C23 - C24	1.382(5)	C29 - C22 - C23	110.4(3)
C23 - C28	1.396(5)	N3 - C22 - C23	112.9(3)
C24 - C25	1.395(6)	C24 - C23 - C28	118.4(4)
C25 - C26	1.369(6)	C24 - C23 - C22	123.2(3)
C26 - C27	1.388(7)	C28 - C23 - C22	118.2(3)
C29 - C34	1.357(6)	C23 - C24 - C25	120.0(4)
C29 - C30	1.401(6)	C26 - C25 - C24	120.6(4)
C30 - C31	1.387(6)	C25 - C26 - C27	119.4(4)
C31 - C32	1.346(9)	C28 - C27 - C26	120.3(4)
C32 - C33	1.382(10)	C27 - C28 - C23	121.3(4)
C33 - C34	1.367(8)	C34 - C29 - C30	118.2(4)
		C34 - C29 - C22	119.8(4)
		C30 - C29 - C22	121.8(3)
		C31 - C30 - C29	119.8(5)
		C32 - C31 - C30	120.1(6)
		C31 - C32 - C33	120.7(5)
		C34 - C33 - C32	119.1(5)
		C29 - C34 - C33	122.1(6)
$[C_{17}H_{22}N_2(3)]^+$ group			
N5 - C35	1.501(5)	C35 - N5 - C38	108.4(3)
N5 - C38	1.508(5)	C35 - N5 - C39	111.6(3)
N5 - C39	1.526(4)	C38 - N5 - C39	109.5(3)
N6 - C37	1.472(5)	C37 - N6 - C36	110.5(3)
N6 - C36	1.480(6)	C36 - C35 - N5	111.5(3)
C35 - C36	1.490(6)	N6 - C36 - C35	112.0(3)

C37 - C38	1.506(6)	N6 - C37 - C38	110.6(4)		
C39 - C40	1.511(5)	C37 - C38 - N5	111.6(3)		
C39 - C46	1.525(5)	C40 - C39 - C46	113.2(3)		
C40 - C41	1.387(6)	C40 - C39 - N5	111.8(3)		
C40 - C45	1.394(6)	C46 - C39 - N5	110.8(3)		
C41 - C42	1.376(7)	C41 - C40 - C45	118.0(4)		
C42 - C43	1.385(8)	C41 - C40 - C39	117.4(4)		
C43 - C44	1.372(7)	C45 - C40 - C39	124.6(4)		
C44 - C45	1.374(6)	C42 - C41 - C40	120.9(4)		
C46 - C47	1.379(5)	C41 - C42 - C43	120.2(5)		
C46 - C51	1.390(5)	C44 - C43 - C42	119.5(5)		
C47 - C48	1.397(6)	C43 - C44 - C45	120.4(5)		
C48 - C49	1.366(8)	C44 - C45 - C40	121.0(4)		
C49 - C50	1.370(7)	C47 - C46 - C51	118.8(4)		
C50 - C51	1.394(6)	C47 - C46 - C39	118.0(3)		
	C51 - C46 - C39	123.1(3)			
	C46 - C47 - C48	120.0(4)			
	C49 - C48 - C47	121.0(4)			
	C48 - C49 - C50	119.3(4)			
	C49 - C50 - C51	120.6(4)			
	C46 - C51 - C50	120.3(4)			
N1 - H1	0.9800	N3 - H3	0.9800	N5 - H5	0.9800
N2 - H2A	0.8900	N4 - H4C	0.8900	N6 - H6A	0.8900
N2 - H2B	0.8900	N4 - H4D	0.8900	N6 - H6B	0.8900
		H2A - N2 - H2B	108.1		
		H4C - N4 - H4D	108.1		
		H6A - N6 - H6B	108.1		

Table 5.
Selected bond lengths (Å) and bond angles (°) in the organic groups of $[C_{17}H_{22}N_2]_3[P_6O_{18}][H_2O]_8$.

It must be noted that: Firstly, all water molecules are not involved in the inorganic layers which may explain their high thermal factors. Such thermal factor values were observed too in others similar structures [20, 21]. Secondly and except O(1 W) and O(2 W), the others water molecules have a static disorder, being split into two fragments with an occupancy rates of 0.5.

4. Conclusion

Along this work, a new $[C_{17}H_{22}N_2]_3[P_6O_{18}][H_2O]_8$ organic cyclohexaphosphate has been successfully synthesized and grown by solution growth method at ambient temperature. According to the X-ray structural results, the crystal structure is governed by hydrogen bonding and intermolecular interactions, resulting infinite inorganic and organic layers. This analysis indicated that the PO_4 tetrahedra possessed a slightly distorted geometry. The P1-O6-P6 angle of 140.09°, depart significantly from the value generally observed in such anions and The P–P–P angles ranging from 103.08 to 114.21° which averages are 107.62°, show large deviations from the ideal value (120°). The title compound was further characterized by FT-IR NMR spectroscopy and DFT calculation.

Conflict of interest

The authors declare that they have no conflict of interest.

Author details

Lamia Khedhiri
Laboratory of Useful Materials, National Institute of Research and Pysico-Chemical Analysis (INRAP), Technopark of Sidi Thabet, Ariana, Tunisie

*Address all correspondence to: lkhedhiri.loulou@yahoo.com

References

[1] Schülke U, Kayser R. Zur thermischen Dehydratisierung von Lithiumdihydrogenphosphat, -hydrogendiphosphat und -cyclophosphat-Hydraten. Zeitschrift für Anorganische und Allgemeine Chemie. 1985;**531**:167-175. DOI: 10.1002/zaac.19855311223

[2] Shi FN, Shen Z, You XZ, Duan CY. $[H_2(_4H_{10}N_2)]_2(H_2PO_4)_4$: hydrothermal synthesis and single crystal structure of an inclusive supramolecular phosphoric salt. Journal of Molecular Structure. 2000;**523**:143-147. DOI: 10.1016/S0022-2860(99)00404-4

[3] Yokotani A, Sasaki T, Yoshida K, Nakai S. Extremely high damage threshold of a new nonlinear crystal L-arginine phosphate and its deuterium compound. Applied Physics Letters. 1989;**55**:2692-2693. DOI: 10.1063/1.101969.

[4] Hamdi A, Khederi L, Rzaigui M. (IUCr) Tetra kis(2-amino-5-chloro pyridinium) di hydrogen cyclo hexa phosphate. Acta Crystallographica. 2014;**E70**:o342-o343

[5] R. Bel Haj Salah, L. Khederi, M. Rzaigui M. (IUCr) Hexa kis (3-chloro-2-methyl anilinium) cyclo hexa phosphate dihydrate. Acta Crystallographica. 2014;**E70**:o61

[6] Salah RBH, Khedhiri L, Nasr CB, Rzaigui M, Lefebvre F. Synthesis, Structure, and Physicochemical Studies of Hexakis (5-Chloro-2,4-dimethoxyanilinium)cyclohexaphosphate Tetrahydrate. Phosphorus, Sulfur, and Silicon. 2010;**185**:595-601. DOI: 10.1080/10426500902870579

[7] Khedhiri L, Salah RBH, Belam W, Rzaigui M. Hexakis(3,4-dichlorobenzyl ammonium) cyclohexaphosphate hexahydrate. Acta Crystallographica. 2007;**E63**:o2269-o2271. DOI: 10.1107/S1600536807012597

[8] Khedhiri L, Akriche S, Al-Deyab SS, Rzaigui M. Bis(3-azoniapentane-1,5-diaminium) cyclohexaphosphate dihydrate: A monoclinic polymorph. Acta Crystallographica. 2012;**E68**:o2038-o2039. DOI: 10.1107/S1600536812025172

[9] Khedhiri L, Selmi A, Rzaigui M. Synthesis and Characterization of a New Organic Cyclohexaphosphate: $[NH_3-(CH_2)_2-NH_2-(CH_2)_2-NH_3]_2P_6O_{18}.3H_2O$ 2014;**2**:179-192. DOI: 10.15640/jcb.v2n2a10

[10] Grappehaus CA, Li M, Gibson ER, Mashuta MS. Hydrogen-bond networks in the mono- and diprotonated cyclic diamine[9]aneN2S. Journal of Chemical Crystallography. 2004;**34**:5. DOI: 10.1023/B:JOCC.0000014681.88460.f4

[11] Thabet H, Bridi M, Joiuni A, Durif A. Characterization of a new organic-cation cyclotetraphosphate $[NH_3(CH_2)_4NH_3]_2P_4O_{12}.9/_2H_2O$. Journal of the Tunisian Chemical Society. 1995;**10**:693-708

[12] Dzyuba ED, Ya Melnikova R, Borisova et EE, Dunets RA. Russian Journal of Inorganic Chemistry. 1978;**23**:1271

[13] Shûlke U, Kayser R. Zur thermischen Dehydratisierung von Lithiumdihydrogenphosphat, -hydrogendiphosphat und – cyclophosphat-Hydraten. Zeitschrift für anorganische und allgemeine Chemie. 1985;**531**:167

[14] Brandenburg K. Diamond version 2.0 impact GbR, Bonn, Germany. 1998

[15] Baur W. The geometry of polyhedral distortions. Predictive relationships for the phosphate group. Acta Crystallographica. 1974;**B30**:1191-1195

[16] Khedhiri L, Jeanneau E, Lefebvre F, Rzaigui M, Nasr CB. Synthesis and Characterization of a New Cyclohexaphosphate, $(C_6H_7ClN)_6P_6O_{18}.0.5(H_2O)$. Journal of Molecular Structure. 2016;**1105**:87-95. DOI: 10.1016/j.molstruc.2015.10.007

[17] Fezai R, Mezni A, Kahlaoui M, Rzaigui M. Synthesis, structural characterization, electrical properties and antioxidant activity of $[p\text{-} (NH_3) C_6H_4NH_3]_3P_6O_{18}{\cdot}6H_2O$. Journal of Molecular Structure. 2016;**1119**:54-63. DOI: 10.1016/j.molstruc.2016.04.051

[18] Khedhiri L, Jeanneau E, Lefebvre F, Rzaigui et M, Nasr CB. Synthesis and Characterization of a new Cyclohexaphosphate, $(C_9H_{14}N)_4(H_3O)_2(P_6O_{18})$. Journal of Chemical Sciences.2016;**128**:1037-1045. DOI: 10.1007/s12039-016-1105-1

[19] Aloui Z, Abid S, Rzaigui M. Phosphorus. Sulfur and Silicon. The Synthesis and Characterization of a New Cobalt Dimethylphenylpiperazinium Cyclotetraphosphate Hexahydrate, $Co[C_{12}H_{19}N_2]_2P_4O_{12}.6H_2O$. 2006;**181**:1-12. DOI: 10.1080/10426500500366988

[20] Fezai R, Khedhiri L, Rzaigui M. Synthesis, crystal structure, NMR characterization, Thermal analysis and Spectroscopic Characteristics of $[2,3\text{-}(CH_3)_2C_6H_3NH_3]_6P_6O_{18}.2H_2O$. Journal of Advances in Chemistry. 2015;**11**:3498-3512. DOI: 10.24297/jac.v11i2.2217

[21] Bagieu-Beucher M, Averbuch-Pouchot MT, Rzaigui M. Crystal chemistry of cyclo-hexaphosphates. XVII. Structure of chromium cyclohexaphosphate henicosahydrate. Acta Crystallographica. 1991;**C47**:1364. DOI: 10.1107/S0108270191001397

Chapter 2

The Cuprate Ln_2CuO_4 (Ln: Rare Earth): Synthesis, Crystallography, and Applications

Basma Marzougui, Amira Marzouki, Youssef Ben Smida and Riadh Marzouki

Abstract

This chapter is concerned with a study of undoped and doped cuprates of the general formula Ln_2CuO_4 (Ln = rare-earth metal) and $Ln_{2-x}M_xCuO_{4\pm\delta}$ (Ln = rare earth and M = Sr, Ba, Ca, Ln', Bi, and 3d metal). The crystal structures of the undoped and doped cuprates having the notations (T, T′, T*, S, and O), significantly depend, however, on the synthetic route. The topotactic synthesis is a specific method, which allows the transformation of the cuprate from the T to T′ structure. The importance of these materials originates from the discovery of the unconventional superconductors of the Ce-doped Ln_2CuO_4. The cuprate materials could function as insulators or semiconductors which are valuable tools in optoelectronic applications. The doped cuprate materials are good ionic conductors and are found useful as electrodes in fuel cell applications. The undoped cuprates reveal high dielectric properties.

Keywords: cuprate, synthesis, crystal structure, superconductivity, ionic conductivity, optical properties

1. Introduction

Perovskite with the general formula ABO_3 is an important structure in solid-state chemistry which has been applied in many fields. From the ABO_3 structure, several approximate structures can be derived, which are equally important and reveal excellent physical and chemical properties. Historically, Perovskite was first depicted by geologist Gustav Rose in 1830 as a particular mineral $CaTiO_3$ calcium titanate [1]. Today, 'Perovskite' refers to a group of compounds with the same crystal structure and similar unit cell parameters. The partial and total substitution of the cationic atoms of the stochiometric ABO_3 allows to obtain several structures with attractive physical and chemical properties [2]. **Figure 1** shows the phases obtained after the modification of the central compound ABO_3.

Horizontally, the diagram shows that phases of layered structure can be formed by the interlacing of motifs (AO) or (BO_2) and motifs (ABO_3). Vertically, it shows the intermediate phases that can be obtained by varying the oxygen content through the oxidation/reduction process [3, 4].

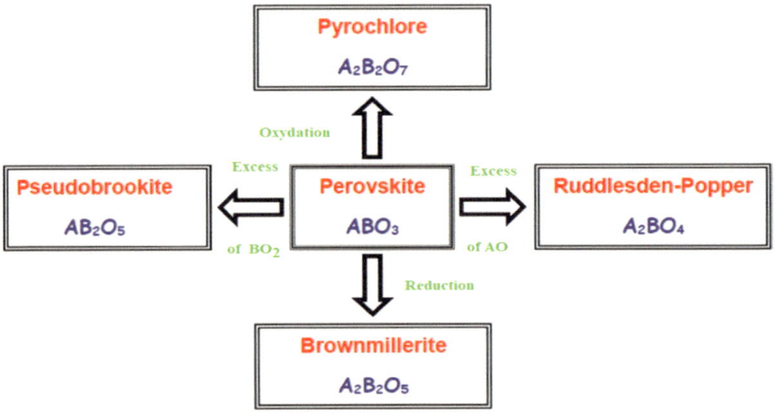

Figure 1.
Derivatives reached from the central structure perovskite ABO_3.

The Ruddlesden–Popper phases (RP), of the general formula $A_{n+1}B_nO_{3n+1}$, have a structure derived from perovskite, which can be described as the stacking of n successive perovskite layers (ABO_3) alternating with one sheet (AO) of NaCl structure along axis c [5].

Like the perovskites, the RP phases show high structural flexibility and more particularly the cuprates of the general formula ACu_mO_n. They are copper-based oxides alloyed with other elements, with different coordination numbers for Cu and consequently different geometry of CuO_2 polyhedra. They may be a chemical compound in which copper forms an anion or complex with an overall negative charge [4].

2. The discovery of the superconductivity of cuprate

Although the undoped cuprates are considered electrical insulators, the "doped" cuprates are regarded as unconventional superconductors [6]. Site A may be rare earth or shared by other rare earth (Sc, Y, and the lanthanides elements) or alkaline earth of different valences (Be, Mg, Sr, Ca, Ba) [7–12]. This gives these materials' different physical properties in relation to the substitution coefficient. These compounds have all different structures but have in common the "active" CuO_2 plans in which the superconductivity is formed [13].

The first cuprate superconductor was discovered in 1986 in lanthanum barium copper oxide by the scientists Georg Bednorz and Karl Alexander Müller [14]. The critical temperature for this material was 35K, much higher than the previous record of 23K [15]. This discovery resulted in a significant increase in research on cuprates, which resulted in hundreds of publications between 1986 and 2001. Bednorz and Müller received the Nobel Prize in Physics in 1987 [16], just one year after their discovery.

Superconductivity in cuprates is considered non-conventional and is not explained by BCS theory (Bardeen-Cooper-Schrieffer) [17]. The potential pairing mechanisms for cuprates superconductivity continue to be the subject of extensive discussion and research. In 1987, Philip Anderson proposed that super exchange could be used as a mechanism for coupling high-temperature superconductors [18]. In 2016, Chinese physicists observed a correlation between a cuprates' critical temperature and the size of the charge transfer gap in that cuprate, offering an explication for the super exchange hypothesis (the strong antiferromagnetic coupling between two neighbor cations through a non-magnetic anion) [19].

3. Crystal structure and synthesis of cuprate

The two typical structural types of Ln_2CuO_4 double oxides are assigned by T (or O) and T′ [16]. The indicative T′ means that the symmetry of the structure is tetragonal while the prefix O designates that the symmetry is orthorhombic. A third structural type exists and is designated by T* for certain oxides of the general formula $Ln_{2-y-x}Ln'_yM_xCuO_4$ (where Ln and Ln′ are lanthanides and M an alkali or alkaline earth such as Ba or Sr) [20–22]. The rare-earth ionic radius of the Ln_2CuO_4 oxides (Ln = lanthanides) is an essential criterion that imposes the type of structure adopted. All Ln_2CuO_4 compounds stabilize in the T′-type structure at room temperature, except for Ln=La which adopts the T-type structure. This difference is due to the enormous size of the La^{3+} ion which prevents the stabilization of the T′ structure. However, the small ionic radii of the other lanthanides favor the crystallization of the T′ phase, the common prototype for this phase of which is the compound Nd_2CuO_4 [23].

3.1 T-type structure (La_2MO_4 system with M = Cu, Ni, and Co)

The T-type structure relates to the La_2MO_4 system where M = Cu, Ni, and Co. At elevated temperatures, the La_2MO_4-type compounds correspond to the K_2NiF_4-type structure of tetragonal symmetry with the space group I4/mmm [24–26]. The structure is described by the stacking, along the c-axis, of sheets of the MO_2 type separated from each other by double layers of the NaCl type or, by the stacking of the layers of octahedron MO_6 of the perovskite-type which are translated relative to each other by (½, ½, and ½) and separated by La-O sheets of the NaCl type [27] (**Figure 2**).

The coordination of the lanthanum atom is equal to 9 or 4 + 4 + 1 oxygen ions with different bond lengths. The shortest bond (strong bond) is between the apical oxygen

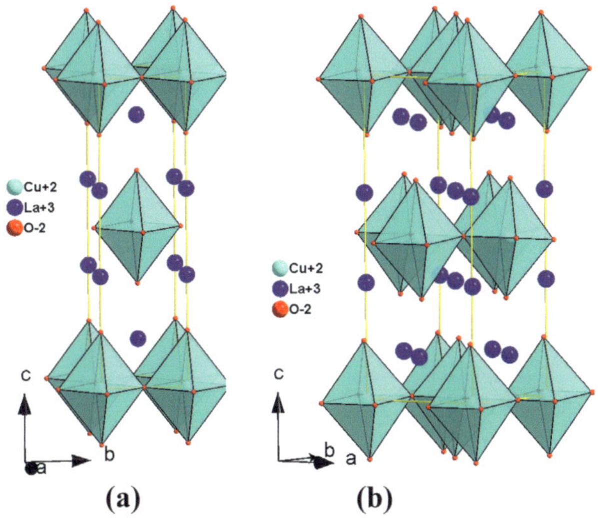

Figure 2.
Unit cell of the T-type structure: (a) HTT quadratic phase (Space Group: I4/mmm) (ICSD 41643), (b) LTO Low-temperature orthorhombic phase (Space Group: Bmab) (ICSD50265).

and the rare-earth atom (Ln–O)⁺. The transition metal is in coordination 4 (four) with respect to the equatorial oxygen and 2 with respect to the apical oxygen. The lengths of Map bonds vary according to the type of transition metal M [28]. Indeed, the Co–Oap and Ni–Oap bonds are equivalent, while the length of the Cu–Oap bond is lengthened on the c-axis by the Jahn Teller effect [29] due to the volume state of the Cu^{2+}. It is equal to 13.15 Å for La_2CuO_4 while it is between 12.55 Å and 12.70 Å for La_2CoO_4 [30] and La_2NiO_4 [31]. The octahedra MO6 is perfectly aligned with the c–axis; in case, HTT (high-temperature tetragonal) is ideal. At room temperature, the phases of the three systems: La_2CoO_4, La_2NiO_4, and La_2NiO_4, have an orthorhombic structure of the Bmab type called "low temperature orthorhombic" (LTO).

3.2 T′-type structure

The structure of the T′ type derives from that of the T type which results from a displacement of the apical oxygen atoms in the tetrahedral sites formed by the lanthanide ions, from where the layers of the NaCl type are replaced by layers of the fluorite type. The coordination number of Cu becomes 4 (square plane coordination) and that of Ln is 8 instead of 9 [32, 33]. This structural transformation leads exclusively to a change in the coordination of the oxide; thus, the cation M is in square plane coordination, and the LnO_2 layers are of the fluorine type and no longer of the NaCl type. Therefore, the crystal structure of the T' phase can be described as an entanglement of fluorite ($Ln/O_2/Ln$) blocks with infinite shell blocks CuO_2. However, the space group remains the same for the 2 structures. The structures of Nd_2CuO_4 [12] and Pr_2CuO_4 [34] are shown in **Figure 3**.

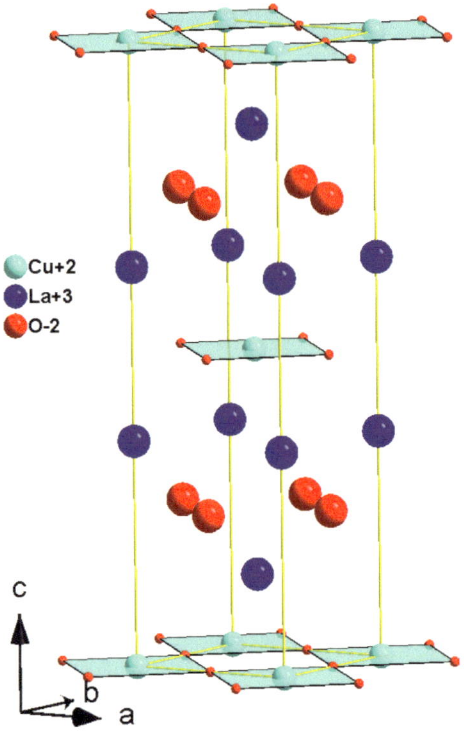

Figure 3.
Structure of the typical quadratic phase T' (ICSD 261660).

	T structure			T' structure	
Atom	**Site**	**Coordinates**	**Atom**	**Site**	**Coordinates**
Ln	4e	(0, 0, z) z ≈ 0,35	Ln	4e	(0, 0, z) z ≈ 0,35
Cu	2a	(0, 0, 0)	**Cu**	2a	(0, 0, 0)
Oeq	4c	(0, 1/2, 0)	**O1**	4c	(0, 1/2, 0)
Oap	4e	(0, 0, z) z ≈ 0,18	**O2**	4d	(0, 1/2/, 1/4)

Table 1.
Atomic positions of the T and T' structures with the space group I4/mmm.

The atomic positions of the two structures T and T' are presented in **Table 1**. The values of the atomic coordinates have been obtained from the crystallographic information file (CIF) obtained from the ICSD database.

3.3 T* type structure

The structure of the T* type consists of a stack of pyramids' layers (CuO_5) separated either by layers (LnO) of the NaCl type, or by layers of the fluorite type (**Figure 4**). As a result, this structure is intermediate between those of the T and T′ types. The stacking sequence of the atomic planes along the c-axis is A | B | A | C | A, with A: CuO_2, B: Ln–O_2–Ln, and C: LnO–LnO. The Cu^{2+} cation is surrounded by five oxygen atoms, thus, forming a square-based pyramid while the Ln^{3+} cations are distributed in the coordination sites 9 and 8 in an ordered manner [22]. The structure of the T* type phase is obtained by following a thermodynamic competition between the factors, e.g., chemical composition and pressure influencing the stability of the T and T′ phases. Due to this competition, the existence of this phase is extremely limited and coincides with the following values of the tolerance factor: $0.85 \leq t \leq 0.86$ [31]. The existence and presence of two ions Ln^{3+} and Ln'^{3+} of varied sizes cause the crystallization of the oxide of type $(Ln, Ln')_2CuO_4$ in the T* type structure. This is the case where Ln is La^{3+} and Ln′ is less voluminous ion such as Sm^{3+}, Eu^{3+} , Gd^{3+} , Dy^{3+}, and Tb^{3+} . The domain of the T*-type phase existence can be increased by the presence of a divalent cation with a large ionic radius such as Sr^{2+} which usually occupies the coordination site 9 [32]. Thus, the structure of the T* phase can include several oxides such as $(La, Ln, Sr)_2CuO_4$ and $(La, Ln')_2CuO_4$ (Ln = Sm, Eu, Gd, or Tb, and Ln′ = Dy, Tb, or Nd) [30]. The structure of the T* type phase has also a quadratic symmetry, and the space group of Nd2–x–yCexSryCuO4- δ is P4/nmm [35] for example.

3.4 S-Type structure

The S-type structure [36] is a model used mainly to describe oxygen-deficient compounds such as Ln2CuO4–x (Ln = Pr, Nd, Sm, Eu, Gd) [37]. Unlike T, T', and T*, the S phase has oxygen vacancies at the equatorial sites, half of which are occupied. The oxygen vacancies are ordered such that Cu adopts a square planar coordination as shown in **Figure 5**. The copper atoms are surrounded by 2Oap and 2Oeq instead of 4Oeq in the T' phase. This is a major difference between the two phases S and T' since the arrangement of CuO4 square planes in the S-type structure does not form 2D layers, but 1D chains which share the following corners and orientations [38, 39]. The S-type phase has orthorhombic symmetry with a space group Immm [39].

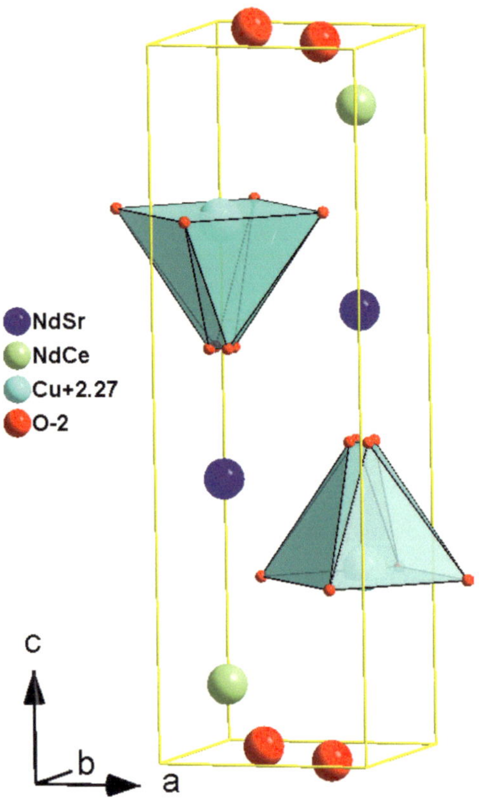

Figure 4.
Projection of the crystal structure of $(Nd_{1.32}Sr_{0.41}Ce_{0.27})CuO_{3.93}$ (ICSD 65871) which belongs to the T structure type.*

4. Effect of doping in phase transition of cuprates

The nature of the Cu–O bond in cuprates is strongly related to doping. Various configurations are found. For example, for the La2–xSrxCuO4 system [40, 41], the structure is of the K2NiF4 type (T phase). It contains CuO6 octahedra arranged in a planar row.

While the Nd2–xCexCuO4 [42] system exhibits a structure similar to Nd2CuO4 type (T' phase) in which the apical oxygens in the T phase structure are shifted away from the Cu atoms in order to form lines of oxygen atoms along the c-axis perpendicular to the CuO2 planes. In the Sr2CuO3 phase (S phase), the CuO_3 chains, running along the a-axis of the orthorhombic structure, were isolated [43]. The T, T′, and S structures are shown in **Figure 6**. Due to the elimination of half of the oxygen atoms noticed in the CuO2 planes, the structure is transformed from the T phase to the S phase. The translation of the apical oxygens in the face positions of the lattice transforms the structure of the T phase to T' phase. The transformation between the phases T, T′, and S is observed in the Nd2CuO4 -Sr2CuO4 system [41]. Indeed, heating T-type phases such as La2CuO4, La2NiO4, and La2CoO4 under various oxygen pressures lead to the formation of oxygen-rich phases, with biphasic regions between these phases and the stoichiometric compounds La2MO4

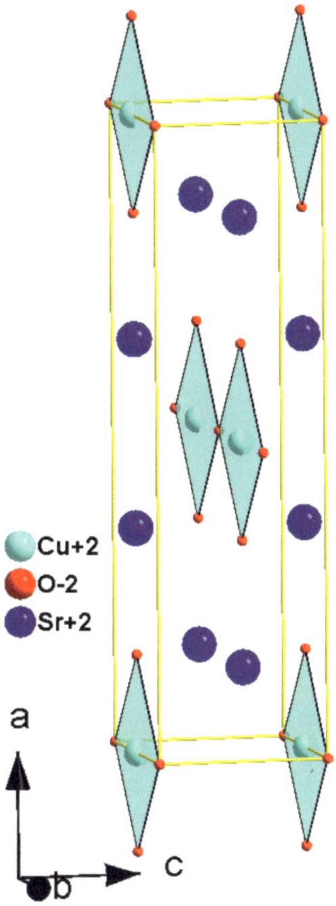

Figure 5.
Projection of the crystal structure of $SrCuO_3$ which belongs to the S structure type (ICSD 15127).

(M: Cu, Ni, Co) [42]. In the case of La2CuO4, the obtained sample is La2CuO4.08 and La2CuO4.03 [44, 45].

The oxygen-rich phase shows superconductivity below 40 K. The reduction of Ln2CuO4 compounds (Ln: La, Pr, Nd, Sm, Eu, Gd) using hydrogen at low temperature [10] leads to the formation of new Ln2CuO4–δ compounds, with δ = 1/3 for Ln = La and δ = ½ for Ln = Pr, Nd, Sm, Eu, and Gd. For the compounds with La, Pr, and Nd, they exhibit structures similar to Sr2CuO3 [46–48].

5. The influence of hole doping on the antiferromagnetic order of CuO_2 planes

Doping is the operation that modifies the concentration of charge carriers in the CuO2 planes. There are two ways of doping the compound either by substituting a cation with another of different valences, for example, in the La2–xSrxCuO4 system, the La^{3+} is substituted by Sr^{2+} [49], or simply by adding oxygen. In the undoped compounds, the planes adjacent to the CuO2 planes consist of trivalent

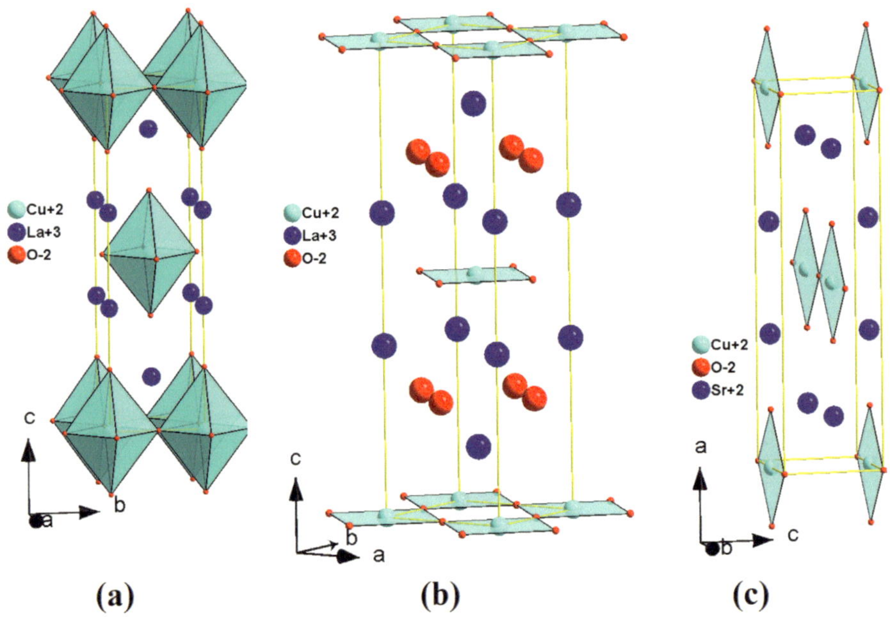

Figure 6.
Structures of (a) T-La2CuO4, (b) T′-Nd2CuO4, and (c) S-Sr2CuO3.

cations X^{3+} (La^{3+} in the phase La2CuO4 or Bi^{3+} for Bi2Sr2CaCu2O8). Only two of the three electrons provided by X^{3+} are needed for bonding X^{3+}-O_2^-. The electron remaining is transferred to blueprints $Cu^{2+}(O_2^-)_2$. The unit cell CuO_2 takes an electron from the two neighboring layers XO, thus, ensuring electronic neutrality. However, when a divalent Z^{2+} ion is partially substituted for the trivalent X^{3+} ion, an electron deficit is created in the CuO_2 planes [50]. This process can also be described as introducing holes in the copper-oxygen planes. These holes transform the $3d^9$ states of Cu^{2+} into Cu^{3+} (S = 0), and these ions represent a Cu^{2+} (S = ½) bonding state with a hole residing mainly in the four neighboring 2p orbitals of the oxygens that is called a “Zhang-Rice” singlet [51]. The introduction of hole-like charge carriers significantly alters the long-range antiferromagnetic ordering of the system as shown in **Figure** 7.

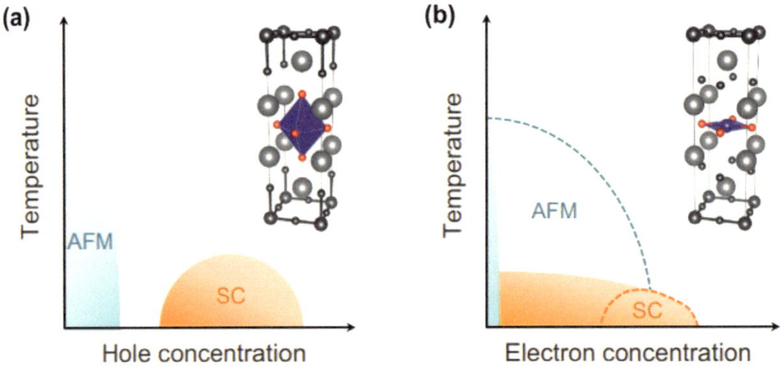

Figure 7.
Schematic representation of the influence of hole doping on the antiferromagnetic order of CuO2 planes [52].

6. Electric properties of cuprate

Undoped Ln2CuO4 cuprates are weak electrical conductors [53] because they have neither free electrons nor anionic vacancy. This low conductivity is due to the presence of two oxidation states of copper Cu^{2+} and Cu^{+}. However, doping gives rise to defects or free electrons which allow electrical conduction.

An electrical conductor allows the passage of electrical current and indicates that in this solid, there are structural defects and/or conduction electrons. For La2–x(Ba,Sr)xCuO4±δ, each electron with a valence of 6s of Ba or 5s of Sr interacts with 2p of oxygen [54]. This interaction corresponds to the energy that participates in the formation of the solid. This energy associated with the interaction stabilizes the valence electron and prevents it from participating in electrical conductivity, which makes the material insulating. It is, therefore, necessary for the valence electrons to be able to participate in electrical conductivity so that they are freed from their bond of valence.

When the electrical current is established, a part of the electrical energy is still distributed over all the ordered atoms, barely modifying the different movements of the electrons, which can no longer leave their atom except in the absence of a defect. On the other hand, for atoms in a defect position, their conduction electron no longer has a synchronous movement with the others, and they can more easily receive electrical energy. As a result, when the electrical current is established, the conduction electrons of the atoms in the defect position can gradually leave their atom, which allows the disturbances generated by the connections to propagate [55]. It is the electrical energy that allows them to cross the gap, which without it, it retains in their corresponding atom. It is appropriate to call such defects with conductors to distinguish them from other resistant structural defects. This is a situation reminiscent of the semiconductor state, the difference coming from the number of defects, and more numerous carriers in a metal or a metallic oxide. Metal is a conductor in which conduction electrons are excited by energy electric.

The mechanism of conductivity in cuprates considers the conductive solid as a mixture of two phases: one ordered and the other disordered. If in the disordered phase, there are too many resisting defects close to the paths of the electrons, then the body is conductive, semi-conductive, or insulating at any temperature [56]. If the number of resistant defects is sufficiently low, the body is superconducting. The notion of resistant defects has the advantage of making it possible to understand the significant difference between oxides and metals. The 2p holes on the oxygen atoms are resistant defects. Due to the importance of electrical properties for this family of materials, they are used in several areas of electricity and energy storage such as solid-oxide fuel cells and capacitors. Indeed, cuprates are used as cathodes for fuel cells. These cathodes must be good electrical and ionic conductors, stable in an oxidizing medium, and compatible with the electrolyte and interconnector materials. The most encountered materials are of the $A_{1-x}Sr_xBO_3$ type, such as manganite and lanthanum cobaltite doped with strontium La0.6Sr0.4MnO3 (LSM) [57] and La0.7Sr0.3CoO3–δ (LSC) [58]. Recently, new materials of general formula A2MO4+δ with RP structure have been studied and show promising results.

7. Dielectric properties

Doped Ln2CuO4 lanthanide cuprates are also characterized by their high dielectric properties due to the high value of the dielectric constant ε′ (ε′ > 104) and the low dielectric loss [59, 60]. The La2CuO4 oxide has a dielectric constant ε′ in the range

103–104 [61]. The dielectric constant of La1.95 Sr0.05CuO4±δ is about 2 × 105 much larger than that of La2CuO4 [11]. The increase in this constant was explained by the increase in the concentration of holes in the material [62]. For the cuprate Eu2CuO4, the dielectric constant was about 103–104 in the frequency range between 1 kHz and 1 MHz for the temperature range from 173 K to 423 K [63].

Indeed, at the microscopic scale, the dielectric permittivity of the material is linked to the electrical polarizability of the molecules or atoms constituting the material. Dielectric materials are polarized in an applied electric field [64]. A certain amount of time is required to orient the dielectric dipoles depending on the direction of the applied electric field. This period is called "relaxation time," which can be attributed to an inhomogeneous microstructure consisting of cuprate grains, separated by insulating grain boundaries.

The dielectric losses encountered in most materials, in particular noticed for cuprate, originate directly from polarization due to orientation because its range of variation can be located between 10^2 and 10^5 Hz [65]. This behavior can be explained in terms that at low frequency, there is a phase shift between the polarization vector and the electric field [64]. But at sufficiently high frequencies, the duration of the electric field is reduced compared to the relaxation time of the permanent dipoles. Thus, the orientation of the latter is no longer influenced by the electric field and remains random. We also observe that the dielectric loss decreases when the temperature increases because the conduction in the dielectric ceases and is believed to be a negligible phenomenon.

8. Optical properties of cuprate

Cuprates represent a very interesting class of semiconductor and superconductor materials, widely known for their important technological applications in the field of display devices, optical smart windows, electrochromic devices (ECDs), and gas sensors [66, 67]. These materials of which optical properties vary with the reduction and/or the oxidation by the injection of ions and electrons which modify their electronic structure are responsible for the change of its visual qualities.

So far, most of the research on Ln2CuO4 and Ln2–xMxCuO4 (Ln = lanthanides, M = Ca^{2+}, Sr^{2+}, and Ba^{2+}) have focused on superconductivity, but few studies in the literature on the preparation and optical properties of Ln2CuO4 and Ln2–xMx-CuO4 materials in the UV–VIS–NIR region are reported [68]. Doped and undoped Ln2CuO4 can be used as photocathodes for the photo-electrochemical decomposition of water, in which the Cu^{2+} ion provides the small bandgap of this material while the incorporated lanthanum ion provides the energy-level adjustment [69]. According to several studies [46], the La2CuO4 electrode acts as a photocathode for the photo-electrochemical decomposition of water and presents a photo-current of 0.5 mA/cm^2. For effective photocatalysts, the bandgap must be large enough to support the 1.23 eV dissociation energy of water. On the other hand, the bandgap should be less than 2.1 eV [70], which would allow the materials to capture and absorb most of the solar energy. For example, La2CuO4 shows broad absorption in the UV–visible region [200–800 nm] with an energy band of about 1.24 eV [71, 72] attributed to $O^{2-} \rightarrow Cu^{2+}$ charge transfer. For Ln2CuO4 (Ln=Pr, Nd, Sm, and Gd), the gap energy is 0.79 eV; 1.06 eV; 1.20 eV, and 1.36 eV, respectively [73]. Chyi-Ching et al. [74] related the observed optical bandgap to the rare-earth ionic radius and show that the bandgap decreases with increasing ionic radii. The optical properties of lanthanum copper oxide-doped La2–xMxCuO4–δ (M = Ca, Sr, and Ba) have been the subject of some articles [71, 75, 76]. They *consider* the effect of the grain size on the optical properties of the prepared samples. Therefore, they used soft chemical synthesis methods, such as sol–gel, combustion and co-precipitation [10, 11, 77]. We can see that the optical bandgap decreases with increasing Ca^{2+} and Ba^{2+} levels but evolves in the same direction in the case of Sr^{2+} and undoped La2CuO4, which was 1.88 eV [78]. This value is different from the value quoted by other authors [79].

Throughout the UV–VIS–NIR region, the La2–xCaxCuO4 cuprates have a wide absorption band and a bandgap that increases linearly with the doping level at $0 \leq x \leq 0.12$; which will respond effectively to its use as a photocatalyst [62, 69].

9. Topotactic reduction of cuprates

Ln2CuO4 compounds (Ln = lanthanide) can also be reduced to lead to Ln2CuO4–δ oxygen-deficient phases. Topochemical deintercalation of O^{2-} ions from transition metal oxides can be used to prepare a large number of metastable phases that contain coordination geometries and transition metal centers in unusual oxidation states [80]. The Ln2CuO4–δ oxygen-deficient phases obtained after reduction constitute an obligatory intermediate step for the transformation of the T to T’ phase [81].

The reduction reaction of a transition metal oxide is normally conducted by artificially reducing the partial pressure of oxygen in the system, either by pumping oxygen or purging with an inert atmosphere (stream of H2 gas) or by the use of highly electropositive binary metal hydrides (LiH, NaH, MgH2, CaH2, SrH2, and BaH2) in a vacuum tube.

The importance of the T’ phase is due to the superconductivity properties of the cuprates. Therefore, numerous attempts have investigated the possibility of transforming the T phase into the T′ phase [10, 80–82] use of a topotactic transformation. For cuprates, e.g., topotactic transformation can be performed in two steps: reduction using hydrides of electropositive metals such as NaH, LiH, and CaH_2 or using H_2. The reduction step creates it possible to find an intermediate phase that is transformed by oxidation to the required structure. For example, using CaH_2, $La_{1.8}Nd_{0.2}CuO_{3.5}$ is synthesized from $La_{1.8}Nd_{0.2}CuO_4$ [10]. T′-$(La\text{-}Ln)_2CuO_4$ phase was first synthesized by Tsukada et al. [83] at a low temperature of 600°C. In this context, researchers are chasing stability conditions to realize whether T′-La_2CuO_4 phase exists only as a thin layer. Thus, using the co-precipitation method, Takayama Muromachi et al. [84] stabilize T′-$La_{1.8}Y_{0.2}CuO_4$ phase at low temperature (600°C).

10. Synthesis of new $Ln_2CuO_{3.5}$ compounds by reduction with CaH_2

New compounds were synthesized using hydrogen reduction as a synthetic method to obtain compounds of the Ln2CuO3.5 (Ln: Pr, Nd, Sm, Eu, Gd) [23, 79, 85] and La2CuO3.67 type [78]. These interesting phases are obtained by heating the compounds of the Ln2CuO4 mother phases under a reducing atmosphere of the order of 5% moles of H2/He at a moderate temperature of 300°C. After inspection with X-ray diffraction, the structures obtained for La, Pr, and Nd are all similar to those of the Sr2CuO3 [86], whereas those of the compounds containing Sm, Eu, and Gd are different. The oxidation of the compound La2CuO3.67 in the temperature range of 300–500°C leads to a new La2CuO4+δ system of the T′ type similar to the structure of Nd_2CuO_4 [82]. The oxygen excess is exceptionally high, ranging from 4.10 to 4.42. Annealing beyond 620°C in the presence of atmospheric oxygen transforms the structure back to the original structure of the K_2NiF_4 (T phase) type, although the excess oxygen varies up to 4.06 and with a small orthorhombicity compared with the starting product La_2CuO_4. The oxidation of the other S-$Ln_2CuO_{3.5}$ phases (Ln: Pr, Nd, Sm, Eu, and Gd) above 300°C leads to the original T′ structure, but the excess oxygen varies between 3.98 for the compounds (Eu, Gd) and 4.03 for the compound Nd [39, 87].

There is a notable change in the lattice parameters between the S, T′, and T phases. For La_2CuO_x, the transition from the T structure to the S structure leads to the elongation of the a- and b-axes and to compression of the c-axis, with respect to the structure T. Consequently, a large increase in the volume of the unit cell (~6%) is observed [88, 89]. The decrease in the c-axis was associated with the loss of oxygen in the CuO_2 plane, as in the example of $La_{2-x}Sr_xCuO_{4-y}$. The transformation of the T′-$La_2CuO_{4+\delta}$ phase to the T phase produces a big decrease of about 4.3% in the unit cell volume.

11. Conclusion

In this chapter, we have described the most common structures of the cuprate materials, and the chemical reaction used to obtain it, especially the topotactic reaction. The cuprate materials with the general formula Ln_2CuO_4 were used particularly in the superconductivity field, but later, the materials have been successfully used in other applications as semiconductors or as anionic conductors, respectively. Thus, the cuprate of the formula Ln_2CuO_4 could be applied in photolysis, photoluminescence, fuel cells, and as dielectric materials.

Acknowledgements

The authors thank the Tunisian Ministry of Higher Education and Scientific Research for the funding of this work within the framework of the laboratory program contract (LR16CNRSM02). The authors extend their appreciation to the Deanship of Scientific Research at King Khalid University, Saudi Arabia.

Abbreviation

ICSD	Inorganic Crystal Structure Data Base

DOI: http://dx.doi.org/10.5772/intechopen.109193

Author details

Basma Marzougui[1], Amira Marzouki[2], Youssef Ben Smida[1] and Riadh Marzouki[3,4,5]

1 Université de Carthage, Centre National de Recherche en Sciences des Matériaux, Laboratoire de Valorisation des Matériaux Utiles, Soliman, Tunisie

2 Laboratory of Signal Image and Energy Mastery, Engineering National Higher, School of Tunis, Tunis, Tunisie

3 Department of Chemistry, College of Science, King Khalid University, Abha, Saudi Arabia

4 Laboratory of Materials, Crystal Chemistry, and Applied Thermodynamics, LR15ES01, Faculty of Sciences of Tunis, University of Tunis El Manar, Tunisia

5 Faculty of Sciences of Sfax, Chemistry Department, University of Sfax, Tunisia

*Address all correspondence to: rmarzouki@kku.edu.sa

References

[1] Katz EA. Perovskite: Name puzzle and German-Russian odyssey of discovery. Helvetica Chimica Acta. 2020;**103**:e2000061

[2] Kim H-S, Lee C-R, Im J-H, Lee K-B, Moehl T, Marchioro A, et al. Lead iodide perovskite sensitized all-solid-state submicron thin film mesoscopic solar cell with efficiency exceeding 9%. Scientific Reports. 2012;**2**:1-7

[3] Kharton V, Marques F, Atkinson A. Transport properties of solid oxide electrolyte ceramics: a brief review. Solid State Ionics. 2004;**174**:135-149

[4] Colville A, Geller S. The crystal structure of brownmillerite, Ca_2FeAlO_5. Acta Crystallographica, Section B: Structural Crystallography and Crystal Chemistry. 1971;**27**:2311-2315

[5] Beznosikov B, Aleksandrov K. Perovskite-like crystals of the Ruddlesden-Popper series. Crystallography Reports. 2000;**45**:792-798

[6] Belich N, Udalova N, Semenova A, Petrov A, Fateev S, Tarasov A, et al. Perovskite puzzle for revolutionary functional materials. Frontiers in Chemistry. 2020;**8**:550625

[7] Özdemir ZG, Kılıç M, Karabul Y, Mısırlıoğlu BS, Çataltepe ÖA, İçelli O. Temperature dependence of dielectric properties and electrical conductivity of EuBa2Ca2Cu3O9-x. In: WITAM-2016. 2016. p. 156. DOI: 10.1016/j.mssp.2017.02.020

[8] Khandale A, Pahune B, Bhoga S, Kumar R, Tomov R. Development of Pr2-xSrxCuO4±δ mixed ion-electron conducting system as cathode for intermediate temperature solid oxide fuel cell. International Journal of Hydrogen Energy. 2019;**44**:15417-15435

[9] Chaudhari V, Khandale A, Bhoga S. Sr-doped Sm2CuO4 cathode for intermediate temperature solid oxide fuel cells. Solid State Ionics. 2014;**268**:140-149

[10] Marzougui B, Houchati MI, Smida YB, Chniba-Boudjada N, Hamzaoui AH. Topotactic transformation of (T-T') La1. 8Nd0. 2CuO4: Synthesis, structure, electrical properties and oxygen diffusion pathways simulation. International Journal of Hydrogen Energy. 2021;**46**:26788-26800

[11] Marzougui B, Houchati MI, Smida YB, Sdiri N, Marzouki R, Hamzaoui AH. Dielectric and optical properties of RE1. 8Sr0. 2CuO4±δ (RE= La, Pr, Nd). International Journal of Electrochemical Science. 2020;**15**:4072-4088

[12] Bo L, Na L, Liping S, Qiang L, Lihua H, Hui Z. Rare-earth elements doped Nd2CuO4 as Cu-based cathode for intermediate-temperature solid oxide fuel cells. Journal of Alloys and Compounds. 2021;**870**:159397

[13] Müller K. The unique properties of superconductivity in cuprates. Journal of Superconductivity and Novel Magnetism. 2014;**27**:2163-2179

[14] Dahiya S, Rani K. Study of high-temperature superconductivity. Physical Sciences. 2022;**1**:17

[15] Johrendt D. Rare earth based superconducting materials. Rare Earth Chemistry. 2022;**2022**:557

[16] Yamamoto T, Kageyama H. Hydride reductions of transition metal oxides. Chemistry Letters. 2013;**42**:946-953

[17] Hirsch J. Superconducting materials: The whole story. Journal of Superconductivity and Novel Magnetism. 2020;**33**:61-68

[18] Geballe T, Kivelson S. Paired insulators and high-temperature superconductors, Pwa90: A lifetime of emergence. World Scientific. 2016;**2016**:127-133

[19] Ruan W, Hu C, Zhao J, Cai P, Peng Y, Ye C, et al. Relationship between the parent charge transfer gap and maximum transition temperature in cuprates. Science Bulletin. 2016;**61**:1826-1832

[20] Fournier P. T′ and infinite-layer electron-doped cuprates. Physica C: Superconductivity and Its Applications. 2015;**514**:314-338

[21] Lv Y-F, Wang W-L, Peng J-P, Ding H, Wang Y, Wang L, et al. Mapping the electronic structure of each ingredient oxide layer of high-T c cuprate superconductor Bi 2 Sr 2 CaCu 2 O 8+ δ. Physical Review Letters. 2015;**115**:237002

[22] Cyr-Choinière O, Daou R, Laliberté F, Collignon C, Badoux S, LeBoeuf D, et al. Pseudogap temperature T* of cuprate superconductors from the Nernst effect. Physical Review B. 2018;**97**:064502

[23] Naito M, Krockenberger Y, Ikeda A, Yamamoto H. Reassessment of the electronic state, magnetism, and superconductivity in high-Tc cuprates with the Nd2CuO4 structure. Physica C: Superconductivity and Its Applications. 2016;**523**:28-54

[24] Yang G, Su C, Ran R, Tade MO, Shao Z. Advanced symmetric solid oxide fuel cell with an infiltrated K2NiF4-type La2NiO4 electrode. Energy & Fuels. 2014;**28**:356-362

[25] Wei HI, Adamo C, Nowadnick EA, Lochocki EB, Chatterjee S, Ruf JP, et al. Electron doping of the parent cuprate La_2CuO_4 without cation substitution. Physical Review Letters. 2016;**117**:147002

[26] Kasem MR, Hoshi K, Jha R, Katsuno M, Yamashita A, Goto Y, et al. Superconducting properties of high-entropy-alloy tellurides M-Te (M: Ag, In, Cd, Sn, Sb, Pb, Bi) with a NaCl-type structure. Applied Physics Express. 2020;**13**:033001

[27] Kataoka R, Taguchi N, Kojima T, Takeichi N, Kiyobayashi T. Improving the oxygen redox stability of NaCl-type cation disordered Li_2MnO_3 in a composite structure of Li_2MnO_3 and spinel-type $LiMn_2O_4$. Journal of Materials Chemistry A. 2019;**7**:5381-5390

[28] Garcia-Barriocanal J, Kobrinskii A, Leng X, Kinney J, Yang B, Snyder S, et al. Electronically driven superconductor-insulator transition in electrostatically doped La_2CuO_{4+} δ thin films. Physical Review B. 2013;**87**:024509

[29] Suyolcu YE, Wang Y, Sigle W, Baiutti F, Cristiani G, Logvenov G, et al. Octahedral distortions at high-temperature superconducting La_2CuO_4 interfaces: Visualizing Jahn–Teller effects. Advanced Materials Interfaces. 2017;**4**:1700737

[30] Wong-Ng W, Laws WJ, Yan Y. Phase diagram and crystal chemistry of the La–Ca–Co–O system. Solid State Sciences. 2013;**17**:107-110

[31] Akbay T, Staykov A, Druce J, Téllez H, Ishihara T, Kilner JA. The interaction of molecular oxygen on LaO terminated surfaces of La_2NiO_4. Journal of Materials Chemistry A. 2016;**4**:13113-13124

[32] Pikalova EY, Sadykov V, Filonova E, Eremeev N, Sadovskaya E, Pikalov S, et al. Structure, oxygen transport properties and electrode performance of Ca-substituted Nd2NiO4. Solid State Ionics. 2019;**335**:53-60

[33] Ishikawa K. Crystal structure of Nd2NiO4. 08. Solid State Ionics. 2014;**262**:682-686

[34] Li H, Cai Z, Li Q, Sun C, Zhao H. Electrochemical investigation of Pr2CuO4-based composite cathode for intermediate-temperature solid oxide fuel cells. Journal of Alloys and Compounds. 2016;**688**:972-977

[35] Fujita M, Suzuki KM, Asano S, Koda A, Okabe H, Kadono R. μSR study of magnetism in the as-prepared and non-superconducting-La0.9Eu0.9Sr0.2CuO4. In: Proceedings of the 14th International Conference on Muon Spin Rotation, Relaxation and Resonance (μSR2017). 2018. p. 011026. DOI: 10.7566/JPSCP.21.011026

[36] Kitagawa S, Higuchi T, Manago M, Yamanaka T, Ishida K, Jeevan H, et al. Magnetic and superconducting properties of an S-type single-crystal $CeCu_2Si_2$ probed by Cu 63 nuclear magnetic resonance and nuclear quadrupole resonance. Physical Review B. 2017;**96**:134506

[37] 蘇玉, High-Pressure Synthesis, Crystal Structure and Physical Properties of Layered Manganese and Zinc Oxyhalides, 北海道大学. 2017. Available from: http://hdl.handle.net/2115/67433

[38] Lenarčič Z, Prelovšek P. Charge recombination in undoped cuprates. Physical Review B. 2014;**90**:235136

[39] Houchati MI, Houchati I, Midouni A, Ceretti M, Hamzaoui A, Paulus W. Topotactic reduction and phase transitions in (T, T′) La1.8Pr0.2CuO4. Arabian Journal of Chemistry. 2017;**10**:S3792-S3797

[40] Wen J-J, Huang H, Lee S-J, Jang H, Knight J, Lee Y, et al. Observation of two types of charge-density-wave orders in superconducting La2-xSrxCuO4. Nature Communications. 2019;**10**:1-6

[41] Tidey JP, Keegan C, Bristowe NC, Mostofi AA, Hong Z-M, Chen B-H, et al. Structural origins of the low-temperature orthorhombic to low-temperature tetragonal phase transition in high-T c cuprates. Physical Review B. 2022;**106**:085112

[42] Neto EH, Comin R, He F, Sutarto R, Jiang Y, Greene RL, et al. Charge ordering in the electron-doped superconductor Nd2-xCexCuO4. 2014. DOI: 10.48550/arXiv.1410.2253

[43] Bobylev I, Naumov S, Zyuzeva N, Telegin S. Study of the conductivity of perovskite-like cuprates Y (Pr) Ba2Cu3Oy and Sr2CuO3. Physics of Metals and Metallography. 2018;**119**:1175-1179

[44] Jorgensen JD, Lightfoot P, Shiyou P, Dabrowski B. Phase Instability at the Maximum Tc in Oxide Superconductors: Phase Separation in La2CuO4+ x, Nd2-xCexCuO4, and La2-xSrxCuO4. In Advances in Superconductivity III: Proceedings of the 3rd International Symposium on Superconductivity (ISS'90), November 6-9, 1990, Sendai (p. 337). Springer Science & Business Media. December 2012

[45] Grenier JC, Arrouy F, Locquet JP, Monroux C, Pouchard M, Villesuzanne A, et al. The Role of the Additional Oxygen Atoms on the Superconducting Properties of. In Phase Separation in Cuprate Superconductors: Proceedings of the second international workshop on "Phase Separation in Cuprate Superconductors" September 4-10, 1993, Cottbus, Germany (p. 236). Springer Science & Business Media. December 2012

[46] Denisova L, Kargin YF, Chumilina L, Podkopaev O, Denisov V. Synthesis of Pr2CuO4 and its heat capacity in the range 364-1064 K. Inorganic Materials. 2014;**50**:1226-1229

[47] Lahmar H, Benamira M, Messaadia L, Hamdi M, Avramova I,

Trari M. Synthesis, physical and photoelectrochemical properties of Gd2CuO4. Journal of Alloys and Compounds. 2020;**816**:152629

[48] Le Dréau L. Phase transitions and oxygen ordering in La2CoO4+ δ and (T, T')-La2CuO4: Single crystal growth and structural studies using synchrotron and neutron diffraction methods. Université Rennes. 2011;**2011**:1

[49] de Sousa JR, Pacobahyba J, Singh M. A theoretical study of the extinction of antiferromagnetic order by holes and dilution in La2− xSrxCu1−zZnzO4. Solid State Communications. 2009;**149**:131-135

[50] Hamad IJ, Manuel LO, Aligia A. Generalized one-band model based on Zhang-Rice singlets for Tetragonal CuO. Physical Review Letters. 2018;**120**:177001

[51] Akpojotor G. Possible propagation of the Zhang–Rice singlet as a probable Cooper channel in the CuO2 planes. Physics Letters A. 2008;**372**:6992-6995

[52] Jang SW, Sakakibara H, Kino H, Kotani T, Kuroki K, Han MJ. Direct theoretical evidence for weaker correlations in electron-doped and Hg-based hole-doped cuprates. Scientific Reports. 2016;**6**:1-7

[53] Zhao H, Li Q, Sun L. Ln2MO4 cathode materials for solid oxide fuel cells. Science China Chemistry. 2011;**54**:898-910

[54] Niemczyk A, Olszewska A, Du Z, Zhang Z, Świerczek K, Zhao H. Assessment of layered La2-x (Sr, Ba) xCuO4-δ oxides as potential cathode materials for SOFCs. International Journal of Hydrogen Energy. 2018;**43**:15492-15504

[55] Lang W, Marksteiner M, Bodea M, Siraj K, Pedarnig J, Kolarova R, et al. Ion beam irradiation of cuprate high-temperature superconductors: Systematic modification of the electrical properties and fabrication of nanopatterns. Nuclear Instruments and Methods in Physics Research Section B: Beam Interactions with Materials and Atoms. 2012;**272**:300-304

[56] Smith M, Paglione J, Walker M, Taillefer L. Origin of anomalous low-temperature downturns in the thermal conductivity of cuprates. Physical Review B. 2005;**71**:014506

[57] Liu Y, Dai H, Deng J, Li X, Wang Y, Arandiyan H, et al. Au/3DOM La0.6Sr0.4MnO3: Highly active nanocatalysts for the oxidation of carbon monoxide and toluene. Journal of Catalysis. 2013;**305**:146-153

[58] He L, Shu Y, Li W, Liu M. Preparation of La0.7Sr0.3CoO3-δ (LSC)@ MnO2 core/shell nanorods as high-performance electrode materials for supercapacitors. Journal of Materials Science: Materials in Electronics. 2019;**30**:17-25

[59] Varma K, Subbanna G, Ramakrishnan T, Rao C. Dielectric properties of glasses prepared by quenching melts of superconducting Bi-Ca-Sr-Cu-O cuprates. Applied Physics Letters. 1989;**55**:75-77

[60] Özdemir ZG, Çataltepe ÖA, Onbaşlı Ü. Impedance and dielectric properties of mercury cuprate at nonsuperconducting state. International Journal of Modern Physics B. 2015;**29**:1550205

[61] Viskadourakis Z, Radulov I, Petrović A, Mukherjee S, Andersen B, Jelbert G, et al. Low-temperature ferroelectric phase and magnetoelectric coupling in underdoped La 2 CuO 4+ x. Physical Review B. 2012;**85**:214502

[62] Sukumar M, Agila M, Sutha A, Ravi V, Al-Enizi AM, Ubaidullah M, et al. Temperature-dependent phase transition: Structural, optical, magnetic and dielectric properties of La2CuO4 perovskite nanoparticles. Journal of Materials Science: Materials in Electronics. 2022:1-13. DOI 10.1007/s10854-022-09301-7

[63] Salame P, Drai R, Prakash O, Kulkarni A. IBLC effect leading to colossal dielectric constant in layered structured Eu2CuO4 ceramic. Ceramics International. 2014;**40**:4491-4498

[64] Al-Gunaid MQ, Saeed AM, Basavarajaiah S. Impact of nano-perovskite La2CuO4 on dc-conduction, opto-electrical sensing and thermal behavior of PVA nanocomposite films. Polymer-Plastics Technology and Materials. 2020;**59**:469-483

[65] El Batouti M, Fetouh H. A facile new modified method for the preparation of a new cerium-doped lanthanium cuperate perovskite energy storage system using nanotechnology. New Journal of Chemistry. 2021;**45**:8506-8515

[66] Schachinger E, Carbotte JP. Oxygen isotope effect on quasiparticle and optical properties in cuprates. Physical Review B. 2010;**81**:014519

[67] Fausti D, Tobey R, Dean N, Kaiser S, Dienst A, Hoffmann MC, et al. Light-induced superconductivity in a stripe-ordered cuprate. Science. 2011;**331**:189-191

[68] Homes C, Hücker M, Li Q, Xu Z, Wen J, Gu G, et al. Determination of the optical properties of La 2− x Ba x CuO 4 for several dopings, including the anomalous x= 1 8 phase. Physical Review B. 2012;**85**:134510

[69] Li Y, Huang J, Cao L, Wu J, Fei J. Optical properties of La2CuO4 and La2− xCaxCuO4 crystallites in UV–vis–NIR region synthesized by sol–gel process. Materials Characterization. 2012;**64**:36-42

[70] Enhessari M, Shaterian M, Esfahani MJ, Motaharian MN. Synthesis, characterization and optical band gap of La2CuO4 nanoparticles. Materials Science in Semiconductor Processing. 2013;**16**:1517-1520

[71] Sukumar M, Kennedy LJ, Vijaya JJ, Al-Najar B, Bououdina M, Mudhana G. Structural, optical, and magnetic properties of Ca2+ doped La2CuO4 perovskite nanoparticles. Vacuum. 2019;**167**:407-415

[72] Shi Y, Lian J, Wei X, Jin K, Song H, Wei M, et al. Study the optical properties of Pr2CuO4 thin films with different Tc0 via spectroscopic ellipsometry. Materials Research Express. 2019;**6**:106416

[73] van't Spijker H, Simon D, Ooms F. Photocatalytic water splitting by means of undoped and doped La2CuO4 photocathodes. International Journal of Hydrogen Energy. 2008;**33**:6414-6419

[74] Hwang C-C, Wu T-Y, Wan J, Tsai J-S. Development of a novel combustion synthesis method for synthesizing of ceramic oxide powders. Materials Science and Engineering B. 2004;**111**:49-56

[75] Hou X, Ren J, Li F, Ma C, Zhang X, Feng H. Research progress of perovskite materials as catalysts. In: IOP Conference Series: Earth and Environmental Science. IOP Publishing; 2019. p. 032020. DOI 10.1088/1755-1315/295/3/032020

[76] Li Y, Huang J, Cao L, Wu J. Optical properties in the UV–Vis–NIR region of La2-xMxCuO4-δ crystallites synthesized by sol–gel process. Journal of the Chinese Ceramic Society. 2012;**40**:432-435

[77] Jemmali M, Marzougui B, Ben Smida Y, Marzouki R, Triki M. Polycrystalline Powder Synthesis Methods, Crystallization. London, UK: IntechOpen; 2021. pp. 1-16

[78] Ji Q, Meng J, Wang H, An T, Meng J, Liu X. Ruddlesden-Popper-based lanthanum cuprate thin film cathodes for solid oxide fuel cells: Effects of doping and structural transformation on the oxygen reduction reaction. International Journal of Hydrogen Energy. 2021;**46**:27173-27182

[79] Pederzolli D, Attfield J. Nd4Cu2O7: A copper (I) oxide with a novel cooperatively distorted T′ Type structure. Journal of Solid State Chemistry. 1998;**136**:137-140

[80] Houchati MI, Ceretti M, Ritter C, Paulus W. From T to T′-La2CuO4 via oxygen vacancy ordered La2CuO3. 5. Chemistry of Materials. 2012;**24**:3811-3815

[81] Houchati MI, Midouni A, Yahya M, Chniba-Boudjada N, Bardeau J-F, Hamzaoui AH. Phase diagram and transition temperatures in the system (T-T') La2-xNdxCuO4 (x≤ 0.5). Inorganic Chemistry Communications. 2021;**132**:108845

[82] Houchati MI. Mobilité des atomes d'oxygène dans quelques oxydes pérovskites non stoichiométriques: synthèse, structure et dynamique de réseau. Rennes. 2011;**2011**:1

[83] Tsukada A, Krockenberger Y, Noda M, Yamamoto H, Manske D, Alff L, et al. New class of T-prime-structure cuprate superconductors. 2005. DOI: 10.1103/PhysRevB.35.8814

[84] Fujimori A, Takayama-Muromachi E, Uchida Y, Okai B. Spectroscopic evidence for strongly correlated electronic states in La-Sr-Cu and Y-Ba-Cu oxides. Physical Review B. 1987;**35**:8814

[85] Corbel G, Attfield J, Hadermann J, Abakumov A, Alekseeva A, Rozova M, et al. Anion rearrangements in fluorinated Nd2CuO3. 5. Chemistry of Materials. 2003;**15**:189-195

[86] Kloß SD, Weidemann ML, Attfield JP. Preparation of bulk-phase nitride perovskite LaReN3 and topotactic reduction to LaNiO2-Type LaReN2. Angewandte Chemie International Edition. 2021;**60**:22260-22264

[87] Botana AS, Norman M. Layered palladates and their relation to nickelates and cuprates. Physical Review Materials. 2018;**2**:104803

[88] Reedijk J, Poeppelmeier K. Comprehensive inorganic chemistry II: from elements to applications. In V1 Main-Group Elem., Incl. Noble Gases V2 Transition Elem., Lanthanides and Actinides V3 Bioinorganic Fundam. and Appl.: Metals in Nat. Living Syst. and Metals in Toxicology and Med. V4 Solid-State Mater., Incl. Ceramics and Minerals V5 Porous Mater. and Nanomaterials V6 Homogeneous Catal. Appl. V7 Surf. Inorganic Chem. and Heterog. Catal. V8 Coord. and Organometallic Chem. V9 Theory and Methods Elsevier Ltd; 2013. pp. 1-7196

[89] Houchati MI, Chniba-Boudjada N, Hamzaoui AH. Investigations on synthesis and electrical properties of the La1.75Nd0.1Ce0.15CuO3.5 prepared by topotactic reduction with CaH2. Journal of Coordination Chemistry. 2020;**73**:1881-1894

Chapter 3

Pharmaceutical Crystals: Development, Optimization, Characterization and Biopharmaceutical Aspects

Agustina Bongioanni, Maria Soledad Bueno, Belén Alejandra Mezzano, Marcela Raquel Longhi and Claudia Garnero

Abstract

In the solid state, the active pharmaceutical ingredients tend to exhibit crystalline order. In this regard, the interest in the crystalline solid state has expanded to include single-component compounds as well as multicomponent systems such as salts, hydrates, solvates, and co-crystals. The study of crystalline behavior is recognized as an essential component of preformulation research in pharmaceutical sciences and industries. The crystalline form can impact the drug properties such as solubility, dissolution rate, stability, hygroscopicity, and toxicity profile. Therefore, each solid form must be appropriately identified and characterized because it will affect the drug formulation, including the pharmacokinetic, pharmacodynamic, and safety properties of the formulation. In this context, this chapter will cover topics such as synthesis approaches (including nucleation and crystallization procedures), crystal polymorphism, solid state characterization techniques and the impact of crystals on physicochemical and biopharmaceutical properties.

Keywords: drugs, crystallization, characterization, polymorphism, biopharmaceutical properties

1. Introduction

Oral drug delivery is the most used route in the pharmaceutical industry. Over 80 percent of drugs are formulated in solid state, not only because of the advantages referred to noninvasive administration and medication adherence but also for reasons of stability from manipulation and storage of unprocessed material to the drug development process [1–3]. Generally, solid drugs are more chemically stable in the solid state than in solution, where degradation occurs more easily [4]. However, despite the oral delivery potential in comparison with other routes, the oral absorption mechanism of drugs is more complex and requires adequate solubility and stability values in the different portions of the gastrointestinal tract, as well as appropriate dissolution profiles [3]. Furthermore, the numerous structural possibilities of a particular active pharmaceutical ingredient (API) in the solid state, as well as several pharmaceutical production parameters, can have a considerable impact on its chemical, physical, and biopharmaceutical properties [1, 5].

In this context, the aim of solid screening is to select the optimal form with the best characteristics for development. This chapter will go through some of the most important aspects of the API study in solid state.

2. Solid forms of pharmaceutical crystals

A drug can exist in different solid forms, including crystalline and amorphous materials, which are classified into single-component systems and multicomponent systems. In addition, each crystal form may crystallize in many different forms, a property known as polymorphism. Polymorphism can be defined as the ability of a molecule to crystallize in multiple crystal structures with identical chemist composition but different molecular packing, and in some cases, also different conformation [6, 7]. Its study is critical in pharmacy because more than 80% of drugs exhibit this phenomenon in which different polymorphs of the same API may have different properties, affecting the viability, safety, shelf life, solubility, dissolution, stability, toxicity, and bioavailability of oral formulations [8].

Single-component systems include anhydrous or non-solvated drugs. These drugs may have multiple polymorph forms, each of which is identified by a different number (roman or arabic). Ritonavir is a historic case involving problems of dissolution and bioavailability in commercialized drugs associated with a polymorphic transformation of form I to form II, which is more stable but less soluble [9]. Other drugs with polymorphic anhydrous forms include sulfathiazole [10], carbamazepine [11], paracetamol [12], fluconazole [13], among others.

On the other hand, the multicomponent crystal systems comprise drug molecules that have different intermolecular interactions with other molecules (guest molecule or coformer) or ion, resulting in the formation of a new solid form without alterations in the covalent chemistry [7, 14, 15]. This group is formed by solvates, hydrates, co-crystals, salts, and a combination of these (like salt hydrates, salt solvates, co-crystal solvates, co-crystal salt, and co-crystal salt solvates) [14, 16–20]. In addition, each multi-component system may have polymorphs [21]. The multicomponent systems can modify different properties of the API without changing its molecular structure, resulting in improved solubility, dissolution, and bioavailability, among others benefits [15].

Solvates are systems in which the drug molecule and the solvent molecule are trapped in the crystalline lattice interacting via hydrogen bonds (mainly). When the solvent molecule is water the system is called hydrate. Crystal hydrates can exist in stoichiometric relations (monohydrate, dihydrate, etc.) or non-stoichiometric relations [22]. The role of the solvate or water molecule can be as a guest or as a stabilizer of the crystal structure [23]. Some examples of solvents used for forming solvates are dimethyl sulfoxide, ethanol, dimethylformamide, ethyl acetate, acetone, and others [24].

Co-crystals are formed by a neutral drug molecule and a co-crystal former in stoichiometry relation [15, 25]. Both molecules reside in the same crystal lattice and are bonding through non-covalent interactions. In particular, hydrogen bonds are especially important in co-crystals but dipolar, π-stacking, van der Waals and halogen-hydrogen interactions may also stabilize the crystalline structure [6, 14, 15].

The selection of co-crystal formers is based on the functional groups of the API, so molecular recognition is favored by using complementary functional groups. Examples of co-crystal formers are acids, amides, carbohydrates, alcohols, and amino acids [26].

Salts are formed by strong ionic interaction of drug molecules with other molecules or atoms ionized, which act like an oppositely charged counterion. Therefore, the ionizable groups of the API limited its formation [23]. In addition, other interactions can act cooperatively to stabilize the crystal form like hydrogen bonding or coordination interactions [14]. The "pKa rule" is an accepted method used in the salt formation that uses the difference between pKa values of API and coformer to predict their behavior. The salt formation is possible if ΔpKa ($\Delta pKa = pKa_{base} - pKa_{acid}$) is less than 3 [15].

In order to guarantee the API solid form present in the bulk material and its pharmaceutical formulations it is critical to evaluate and characterize each solid form, this topic will be detailed in Section 4.

3. Mechanism of synthesis

As mentioned in the previous section, there exist numerous types of pharmaceutical solid forms in the crystalline state (polymorph, solvate/hydrate, co-crystal, salt). Crystals are solids with a regular array of atoms and molecules built from a translational repetition of the basic structure denominated by unit cell. Thereby, a complete description of the concept of crystallization is fundamental.

Crystallization is a phenomenon that occurs as a result of two different processes. The first is called nucleation, which is the beginning of a phase transition from a supersaturated state that gives rise to the appearance of a small nucleus in a second phase. The second one is the crystal growth process, which involves the evolution layer by layer to determine the crystal packing of the unit cell [27]. The strength of the intermolecular interactions within the unit cell is what determines which layers dominate the crystal growth process [28]. Therefore, the crystals of an API can differ in size relative to the growth of particular faces and the number and type of faces present, ergo they can have different crystal habits, which characterizes the crystal shape (acicular, prismatic, pyramidal, tabular, columnar or lamellar type).

Crystallization is a process of transformation from a solution or melt to the crystalline state. The generation of crystal nuclei is controlled by the crystallization conditions (e.g., solvent, temperature, and supersaturations). Moreover, a solvent or additive in the process of growth may cause competition for a site at an incoming point associated with the layer-by-layer growth process that would be capable of disrupting the magnitude of the intermolecular interactions generating inhibition or interference in the growth directions which is manifest as a change in the overall morphology of the crystal. In industrial crystallization, seeding the supersaturated solution with crystalline material is a common strategy for ensuring batch-to-batch reproducibility and optimizing process robustness by controlling the whole crystallization process by minimizing spontaneous nucleation. In particular, the addition of desired form seeds is the technique most used to control polymorphism [27].

The crystallization process was described using a variety of methodologies (summarized in **Figure 1**), each with its own characteristic, including crystallization from a single solvent, evaporation from a binary mixture of solvents, antisolvent addition, temperature gradient, vapor diffusion, slurrying, and liquid assisted grinding [4, 6, 15].

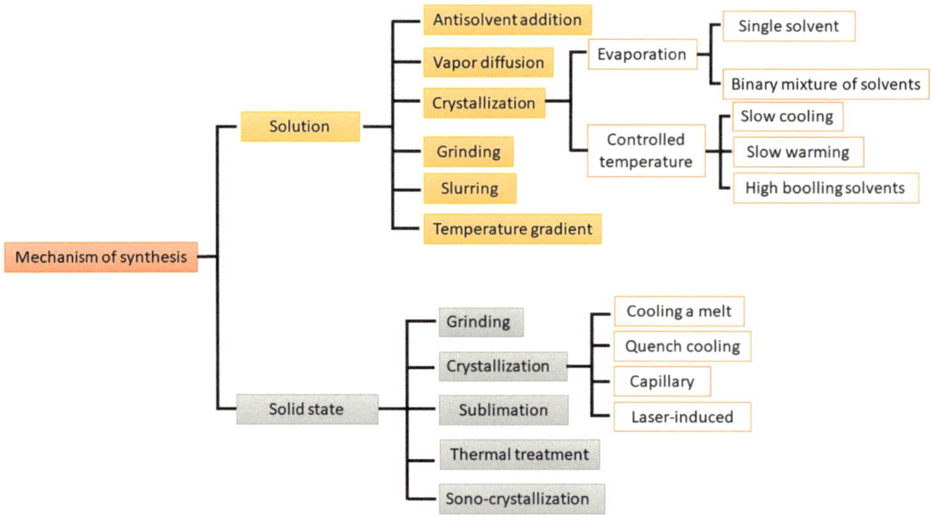

Figure 1.
Methodologies for manufacturing solid pharmaceutical materials.

The crystallization from a solution could proceed in different ways, including slow cooling of a hot saturated solution, slow warming, or by heating the solution to boiling and then quenching cool using an ice bath. On the other hand, if crystallization from a solution is not possible, there are a number of processes that do not require the use of a solvent such as sublimation, thermal treatment, crystallization from the melt, neat grinding, capillary crystallization, laser-induced crystallization and sono-crystallization [27, 29].

The techniques applied for the preparation can use thermodynamic or kinetic conditions, depending on whether the thermodynamic equilibrium is maintained or the situation moves away from equilibrium, respectively, to obtain the crystallization of different crystal forms. The synthesis mechanisms that obtain thermodynamic conditions include slow evaporation, and slow cooling, among others. While kinetic conditions refer to high supersaturation degree, quench cooling, and rapid solvent evaporation, among others [30]. Under stress situations, crystallization kinetics will control the crystal shape, rather than thermodynamics conditions, and the production of more unstable solid forms will be favored kinetically [6].

Additionally, the initial phases of crystallization, determined by the time between supersaturation and the development of nuclei, are critical in regulating the characteristics of the final solid phase, such as purity, crystal structure, and particle size [27]. In general, the most thermodynamically stable crystalline form is preferable. Crystallization performed in close proximity to equilibrium are likely to generate forms of relatively stable or ground-state polymorphs. Though the production of amorphous or metastable forms with increased solubility and dissolution rates may be favored by bioavailability requirements [6]. For example, techniques that produce an abrupt change in the system, such as sublimation or crystallization from the melt, result in a metastable solid form. Thermal desolvation of crystalline solvates can generate amorphous materials, with the solvent contributing to stabilizing the lattice. While techniques such as quench cooling can also be used to obtain amorphous forms. However, these high-energy forms tend to be transformed into a stable form through a solid-solid physical transition, a phase transformation with solvent mediation, or both.

4. Solid state characterization techniques

As previously mentioned, changes in the properties of solids can occur during pharmaceutical drug manufacturing and storage. Normally, drugs are manufactured in a stable crystalline form because the risk of solid state transformations during storage is minimized. However, when developing a solid crystalline form, a rigorous control must be made to determine if the crystalline form is maintained or if there were changes during its production [31].

Currently, there are a variety of techniques to characterize a crystal. Characterization techniques are valuable tools that make it possible to determine the structure, chemical composition, and different properties of a pharmaceutical sample. However, simply one technique will not be able to offer complete information for a solid substance. It is vital to utilize them in a complementary manner in order to acquire acceptable outcomes (**Figure 2**). The most important techniques used in the pharmaceutical field for crystal characterization are those described below.

4.1 Thermal analysis techniques

Thermal analytical methods, which offer information on the thermal behavior of materials, are widely utilized in the physical characterization of pharmaceutical solids. When a sample is exposed to a temperature increase, the observed changes in characteristics can be measured [32, 33]. The thermal methods most commonly applied in the analysis of pharmaceutical solids are differential scanning calorimetry and thermogravimetric analysis.

4.1.1 Differential scanning calorimetry (DSC)

This technique provides information about the physical and energetic properties of a substance subjected to temperature variations. To obtain a DSC thermogram, the

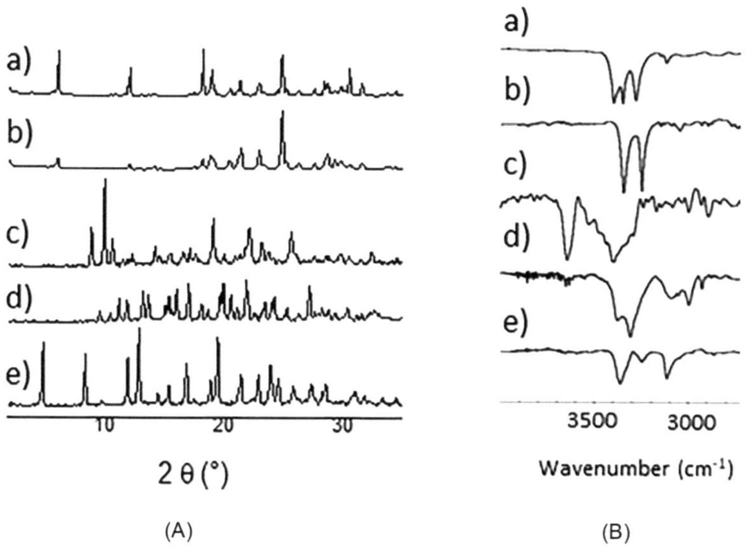

Figure 2.
A) Powder X-ray diffraction patterns and B) Fourier-transform infrared spectra of a) furosemide form I, b) furosemide form II, c) oxytetracycline hydrochloride form I, d) oxytetracycline hydrochloride form II, and e) oxytetracycline hydrochloride form III.

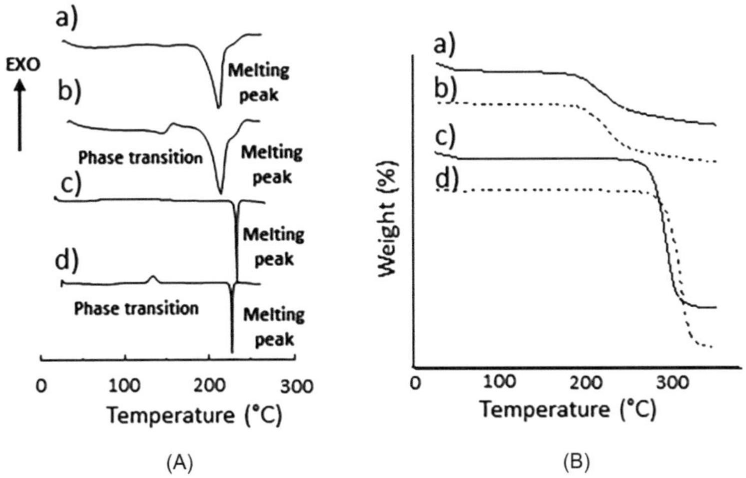

Figure 3.
A) Differential scanning calorimetry curves of a) albendazole form I, b) albendazole form II, c) clarithromycin form II, d) clarithromycin form 0 and B) thermogravimetric curves of a) oxytetracycline hydrochloride form I, b) oxytetracycline hydrochloride form II, c) clarithromycin form 0, d) clarithromycin form II.

difference in heat flux of a sample as a function of temperature or time is measured. Typically, this study is performed by comparing the thermogram of a sample of interest, such as a crystal, to that of a reference sample. The deviation in the thermogram below the reference corresponds to an endothermic transition, while the deviation above the reference relates to an exothermic transition [34]. The curve obtained can be used to determine enthalpies of crystal fusion, phase transition temperatures, purity, degree of crystallinity, type of interaction between molecules, and thermal stability [34, 35].

Numerous studies reported the most popular application of DSC in the field of pharmaceuticals. Their application in the characterization of polymorphic drugs such as albendazole and clarithromycin can be seen in **Figure 3A**. The representative DSC curve for albendazole form I exhibits a fusion endotherm, whereas the profile for form II shows a preceding endo-exothermic event that indicates its polymorphic transformation to form I, followed by an endotherm attributed to form I melting, and lastly an exotherm of decomposition. Similarly, the profile for clarithromycin form 0 exhibits an exotherm corresponding to a solid phase transition, followed by an endotherm assigned to the form II melting event due to its coincidence with the melting endotherm evidenced in the clarithromycin form II DSC curve.

4.1.2 Thermal gravimetric analysis (TGA)

TGA measures the mass change of a sample as a function of temperature or heating time. It is a simple technique that requires a smaller sample size [36]. A thermogravimetric curve shows the mass change due to physical and chemical phenomena such as absorption, melting, sublimation, vaporization, oxidation, reduction, and decomposition events [32].

It is a useful tool to quantify different processes such as crystalline melting, sublimation, or decomposition of a sample, and to elucidate the degree of purity of the API [37]. On the other hand, it is possible to elucidate on the curves, whether the crystals under study contain water or a solvent [36]. Moreover, it allows the detection of solvent loss in a crystal. For instance, information on dehydration/desolvation events for clarithromycin and oxytetracycline hydrochloride polymorphs was obtained (**Figure 3B**), as demonstrated by a mass loss at low temperatures in the TGA profiles. When the TGA profiles of Clarithromycin form 0 and form II are compared, it is clear that form 0 is a solvate, as demonstrated by the weight loss caused by the ethanol evaporation process. On the other hand, TGA profiles of oxytetracycline hydrochloride form I showed a larger mass loss until 100°C than those of form II, indicating variations in solid dehydration. These TGA curves also revealed that form II had higher thermal stability.

4.2 X-ray diffraction

The technique most commonly utilized for identifying and characterizing crystalline materials is X-ray diffraction. Differentiating between crystalline and amorphous forms, identifying distinct solid forms of crystals, defining the crystalline structure of the API, and analyzing the differences between different crystal forms are some of its applications. As a result, it is commonly used in the pharmaceutical field [36]. For example, X-ray diffraction experiments have provided an unequivocal identification of furosemide and oxytetracycline hydrochloride polymorphs (**Figure 2A**), which exhibited clear differences in terms of reflection positions and relative intensity.

Single crystal X-ray diffraction is employed to determine the molecular structure of pharmaceutical materials that exist as single crystals [7, 38, 39]. A three-dimensional picture of the molecule and geometrical properties data in the solid state can be produced by studying a perfectly crystalline sample [40]. Powder X-ray diffraction is applied when the crystalline material is found as a fine-grained powder, rather than a single crystal [7, 38, 39, 41].

4.3 Vibrational spectroscopic techniques

Vibrational spectroscopic techniques are widely used in the pharmaceutical field to identify crystalline solids due are fast, non-destructive, and can characterize solid samples with minimal or no preparation. The most commonly used methods for analyzing crystalline samples are Fourier-transform infrared (FT-IR) and Raman spectroscopy [7, 31].

These techniques are extensively utilized in the study of pharmaceutical solids to characterize amorphous and crystalline phases, identify the structure and composition of different pharmaceutical solid forms, determine the compatibility of mixtures, and establish molecular interactions [42].

For instance, FT-IR spectroscopy has been used in several studies to identify the individual polymorphic forms of a drug confirming that they are structurally distinct. Differences in characteristic FT-IR bands assigned to sulphonamide NH and secondary amine NH stretches were identified between furosemide polymorphs (**Figure 2B**). In the same way, significant differences between oxytetracycline hydrochloride polymorphs were observed in the bands attributed to the OH and amide NH stretching vibrations (**Figure 2B**).

4.4 Solid state nuclear magnetic resonance (ssNMR)

ssNMR is a non-destructive and multinuclear technique that exploits the magnetic properties of certain nuclei, for example, ^{1}H, ^{13}C, ^{15}N, ^{17}O, and ^{19}F. Although it is a non-routinary expensive methodology that has extensive experimental times and robust expertise users are required, it is widely used in pharmaceutical applications [32].

This technique is used to analyze crystalline and amorphous pharmaceutical samples qualitatively and quantitatively, as well as to characterize both APIs and formulations. Structural or dynamic information is obtained from mono and bidimensional experiments based on different nuclear interactions. Their pharmaceutical applications included identification, characterization, and quantitation of different solid forms of an API in bulk samples; determination of conformational and crystalline packing behavior, intra- and intermolecular interactions, internuclear distances; study of amorphous phase properties, stability of API forms, the effects of drug processing, molecular motions, chemical and physical interactions between API-excipient and excipient-excipient, solid state chemical reactivity; and identification of contaminants or degradation products, among others [7, 43, 44].

4.5 Microscopy

Microscopy is considered a tool of great interest in the pharmaceutical field, which is mainly used to examine shape and size and to identify the solid state form in the sample. Different types of microscopes are currently used for the characterization of pharmaceutical crystals [32, 45]. The most relevant are described below.

4.5.1 Scanning electron microscopy (SEM)

SEM is a very useful and versatile tool in the pharmaceutical field. It provides quantitative as well as qualitative information such as morphology, size, size distribution, crystal shape, and consistency of powders or compressed dosage forms by analyzing the images obtained by microscopy. In addition, it allows studying the effects of any interaction with its environment [7, 45].

Microscopic analysis of pharmaceutical crystals using SEM microphotographs reveals significant morphological differences between solids produced using distinct crystallization techniques, allowing each polymorphic form of the drug to be identified. In **Figure 4**, for example, significant differences in particle size and shape can be observed. Albendazole form I appear as small and irregular particles with a predisposition to aggregate while in contrast albendazole form II exhibits self-agglomerate lamellar particles with a smooth surface. Furosemide form I presented hexagonal and tubular compact crystals with a defined surface while furosemide form II shows fine and elongated prism particles. On the other hand, a compact structure with small particles adhered to the surface is observed for norfloxacin form B_I, while the norfloxacin form C crystals are typical hexagonal-like faceted, compact, and with well-defined smooth structures. Finally, oxytetracycline hydrochloride form I have particles with a smooth surface and well-defined edges, form II crystals show compact particles with an irregular surface, form III presents rod-shaped crystals with a smooth surface with defined edges, while the form IV appeared as thin agglomerated needles.

4.5.2 Optical microscopy and polarized light microscopy

An optical microscope is used to observe crystals directly providing information on particle size and shape. In addition, the nucleation events can be visualized by monitoring within situ cameras [32].

Figure 4.
SEM images of the morphology of a) abendazole form I, b) albendazole form II, c) furosemide form I, d) furosemide form II, e) norfloxacin form B_1, f) norfloxacin form C, g) oxytetracycline hydrochloride form I, h) oxytetracycline hydrochloride form III, i) oxytetracycline hydrochloride form II, and j) oxytetracycline hydrochloride form IV.

The utilization of polarized light, on the other hand, optimizes the utility of optical microscopes. Using polarized light microscopy, the interior structure of crystals can be analyzed and determined if the sample is amorphous or crystalline. Due to birefringence, several colors can be seen in a crystalline particle when viewed through crossed polarizers [7, 32, 45].

For example, significant differences in particle size and shape of sulphathiazole precipitated from a variety of solvents and techniques between them and compared

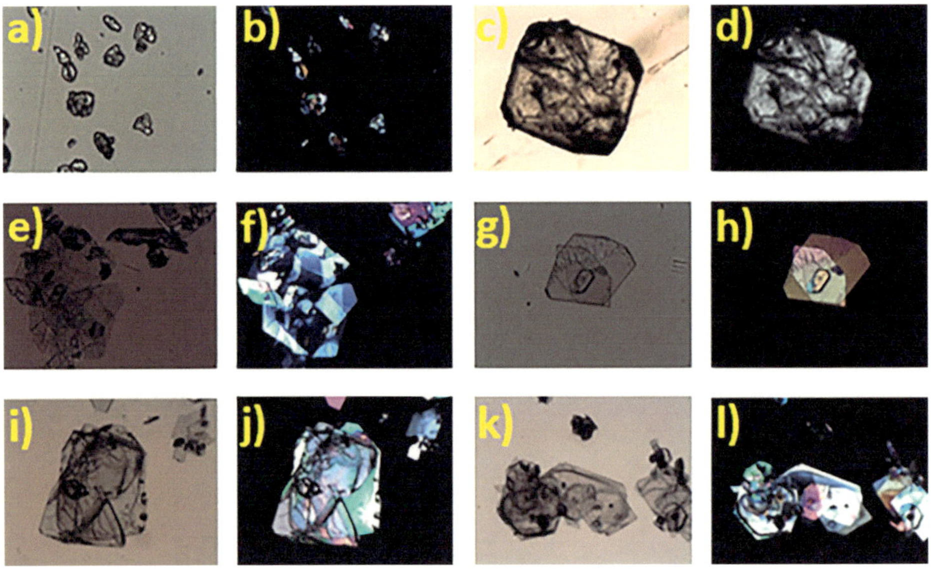

Figure 5.
a), c), e), g), i) and k) images obtained using an optical microscope. b), d), f), h), j) and l) images obtained using a polarized light microscope.

additionally with commercial sulphathiazole can be observed by an optical microscope. In addition, these different sulphathiazole crystal forms exhibit different birefringence under a polarized light microscope. **Figure 5** shows images of commercial sulphathiazole (**Figure 5a** and **b**) and samples obtained by the crystallization of commercial sulphathiazole from methanol heating the solution below the boiling point to the total solution (**Figure 5c** and **d**), from aqueous solution heating below the boiling point to total solution and immediately cooled at freezer temperature (**Figure 5e** and **f**), and from the saturated aqueous solution obtained below 80°C that was exposed to a temperature ramp of 90 to 25°C for one hour and then kept at 25°C for 24 hours (**Figure 5g** and **h**), 1hour (**Figure 5i** and **j**) and 30 minutes (**Figure 5k** and **l**).

5. Solid form impact on physicochemical and biopharmaceutical properties

As described above, the crystallization process determines different habit crystals associated with particular lattice energy, resulting in measurable differences in physical properties. Therefore, different crystalline forms of a drug can be used in pharmaceutical science to improve their physicochemical and biopharmaceutical properties such as melting point, hygroscopicity, solubility, dissolution rate, stability (physical and chemical) mechanical and optical properties. The wettability of an API, for example, has an impact on its solubilization and dissolution processes. The absorption rate of many poorly soluble drugs is determined by their dissolution rate. Hence, drug molecules with poor solubility may lead to slow dissolution in biological fluids, resulting in an erratic bioavailability and consequent sub-optimal efficacy when delivered via the oral route [28, 30, 46].

The crystal morphology of solid drugs influences their dissolving rate due to critical factors such as surface area, size, and even the polymorphic form of the material, which may have a potential impact on the rate and extent of drug absorption.

The size of drug particles and their ability to be wetted by gastrointestinal fluids determine the drug surface area accessible for dissolution. The particle size is dependent on the crystallization conditions or on milling procedures. Therefore, controlled crystallization methods must be used to produce powders with high purity and predetermined particle size distribution for API administration.

The effect of crystal form on the dissolution and bioavailability of the API has been demonstrated. The kinetic transformation and growth conditions in crystallization have a direct effect to generate a particular architecture, which can be a stable polymorph or a metastable form. The polymorph selection process requires a high level of manipulation and control to obtain specific crystal structures grown in selected solvents.

In general, different polymorphs show solubility differences typically smaller than 10 times due to relative differences in free energy. Some examples include the evaluation of solubility, in aqueous and buffer solutions, of several forms of furosemide [47], norfloxacin [48–50], albendazole [51], and oxytetracycline hydrochloride [52]. These studies demonstrate that the molecular arrangement of each polymorphic form and its degree of ionization has a considerable impact on drug solubility (**Table 1**).

Additionally, the effect of polymorphs on bioavailability has a direct impact on pharmacokinetic parameters. A typical example is chloramphenicol palmitate, which exists in 4 solid forms: A, B, C and amorphous structure. Form A is the most stable, however, only the metastable form B and the amorphous solid have biological activity. Aguiar [53] reported that the blood serum level of form B is substantially higher than form A, by nearly an order of magnitude, after oral administration of suspensions at the same dose. It was concluded that form B has high free energy, then is more soluble and thus has a higher rate of absorption and bioavailability.

In some situations, occasionally metastable crystalline or amorphous forms are utilized for drugs orally administered if a faster dissolution rate or higher concentration is desired, in order to achieve rapid absorption and therapeutic effectiveness.

Solution	Furosemide (ug/mL)[a]	Norfloxacin (mg/mL)[a]	Albendazole (ug/mL)[b]	Oxytetracycline hydrochloride (mg/mL)[b]
Aqueous	Form I: 35.6 Form II: 27.4	Form A: 0.32 Form BI: 0.30 Form C: 0.29	Form I: 1.2 Form II: 2.8	Form I: 91 Form II: 6.8 Form III: 6.7
Simulated gastric fluid	Form I: 4.4 Form II: 6.1		Form I: 183 Form II: 320	Form I: 93 Form II: FS Form III: FS
Simulated intestinal fluid			Form I: 3.3 Form II: 7.1	Form I: 76 Form II: 1.20 Form III: 0.91
Buffer pH 6.0		Form A: 0.39 Form BI:1.40 Form C: 2.61		
Buffer pH 8.0		Form A: 0.18 Form BI: 0.60 Form C: 1.38		

[a]*Solubility at 25.0 ± 0.1°C.*
[b]*Solubility at 37.0 ± 0.1°C, FS: freely soluble.*

Table 1.
Influence of crystal form on drug solubility.

Although metastable polymorphs can improve solubility, dissolution, and bioavailability, they can also be transformed into a more thermodynamically stable form during manufacture and storage, which results in unacceptable bioavailability and limits their potential performance. As an example, the polymorphic transformation of chlorpropamide caused by the mechanical energy of tableting compression was described. The heat generated by the compaction process would accelerate the transformation process of the metastable form C into the stable form A, in consequence, its dissolution rate decreases after compression [54, 55].

Furthermore, physical form stability in the gastrointestinal environment should also be considered. During gastrointestinal transit, the transformation in the most stable form is a relevant factor to consider. If the conversion occurs in the course of oral administration, a less soluble form will precipitate reducing oral absorption. For example, Kobayashi [56] demonstrated some differences in the oral pharmacokinetics and bioavailability between carbamazepine polymorphs (anhydrous forms I and III) and the dihydrate form. The plasma concentration-time profiles of polymorphs and dihydrate form differ in correlation with their dissolution profiles, which were in the order form III \form I\dihydrate; furthermore, form III was transformed in situ to dihydrate form faster than form I. By comparing the in vivo performance of carbamazepine at high doses, the form I provide better pharmacokinetic parameters than the other two forms. The inconsistency between the order of initial dissolution rates and pharmacokinetics values suggested a probable rapid transformation of form III to the dihydrate form in the gastrointestinal fluids, resulting in a slowing of dissolution due to the production of the dihydrate form.

A substantial solubility difference between amorphous and crystalline API is observed. The high-energy amorphous solids significantly enhance the solubility of poorly soluble drugs as compared to crystalline forms, resulting in a faster dissolution rate and subsequent oral absorption, which are linked to their metastable nature. In the dry state, the amorphous solid is typically more reactive than the crystalline form due to its higher thermodynamic activity. Furthermore, if exposed to humid conditions, amorphous solids become more hygroscopic, and the absorbed moisture works as a plasticizer, resulting in a substantial increase in molecular mobility. As a result, the chemical stability of an amorphous material is significantly lower than that of the crystalline phase when exposed to moisture. On the other hand, these metastable phases are susceptible to phase transformation during storage, which limits their application in pharmaceutical dosage forms. Although physical and chemical stability of amorphous phases is a major concern, if these high-energy forms can be stabilized to prevent crystallization over their intended storage life using excipients of conventional solid dosage formulation, these solids can be a useful tool for increasing API dissolution in biological fluids given bioavailability enhancement. For example, Yang [57] compared the bioavailability of amorphous and crystalline itraconazole nanoparticles administered via pulmonary. It was observed that amorphous nanoparticulate itraconazole had a rapid dissolution that produced a significantly higher systemic bioavailability than crystalline nanoparticles due to its supersaturation 4.7-times larger, which increased the drug permeation and will be thus beneficial for both local and systemic therapy.

Crystal engineering of co-crystals is another alternative formulation for improving drug attributes including solubility, dissolution, bioavailability, and physical stability of poorly soluble API. The modification of the physicochemical properties of the API and bulk material while maintaining the intrinsic activity of the drug molecule is enabled by co-crystallization during dosage form design [28, 58]. Several examples of API co-crystals with pharmaceutically approved coformers can be found in the literature. Screening for obtaining the optimal solid form is critical in co-crystal development, given the risks and high development industrial costs, because not every co-crystal may significantly enhance the solubility and dissolution rate of the API. A typical example is the carbamazepine-nicotinamide co-crystal that spontaneously converts to carbamazepine dehydrate during dissolution, which has a lower solubility, and the theoretical solubility/dissolution improvement of the co-crystal cannot be obtained [59]. Similarly, co-crystals of efavirenz were developed using several coformers [60]. When compared to pure efavirenz, efavirenz-DL-alanine and efavirenz-oxalic acid co-crystals had higher solubility and enhanced dissolution profiles, while efavirenz-maleic acid and efavirenz-nicotinamide co-crystals had decreased dissolution.

The salt formation has been used to improve the bioavailability since the solubilities of salts are typically higher. Changing the counterions in a salt varies its solubility and dissolution rate, affecting bioavailability, pharmacokinetic profile, and potential toxicity. Also, the salt will impact the chemical stability. Different microenvironmental pH and different molecular patterns in a specific lattice are factors that contribute to the difference between salt and its unionized form or between different salts. Recently, multi-drug salts of norfloxacin have been obtained with diclofenac, diflunisal, and mefenamic acid, as well as norfloxacin salt hydrate with indomethacin. Among them, norfloxacin salts with diflunisal and indomethacin showed higher solubility and permeability and hence increased bioavailability [61].

6. Conclusion

Understanding the characteristics of APIs in the solid state is critical in the field of pharmaceutical sciences since it is the basis for controlling the pharmaceutical performance of final formulations. In order to obtain solid pharmaceutical materials with improved properties, the crystal engineering strategy is used. Different crystallization processes are the experimental key to the solid form screening aiming to select the suitable physical form of a drug. As discussed above, the change in crystal form may not only affect the stability and mechanical attributes of the solid but, more importantly, may compromise the drug absorption through a change in solubility. In practice, is desirable that the drug's physical form does not change during its manufacture and storage life to prevent a significant impact on its quality and bioavailability. Therefore, the characterization of the API solid phases such as polymorphs, solvates, hydrates, salts, co-crystals, and amorphous forms is critical in the early stages of the solid form development as well as a tool for evaluating the influences of manufacturing processes and storage on phase transitions as important factors for product quality assurance.

Conflict of interest

The authors declare no conflict of interest.

Author details

Agustina Bongioanni[†], Maria Soledad Bueno[†], Belén Alejandra Mezzano[†], Marcela Raquel Longhi and Claudia Garnero*
Department of Pharmaceutical Sciences, Faculty of Chemical Sciences National University of Córdoba, Unit for Research and Development in Pharmaceutical Technology, (UNITEFA) CONICET, Córdoba, Argentina

*Address all correspondence to: cgarnero@unc.edu.ar

† These authors contributed equally to this work.

References

[1] Couillaud BM, Espeau P, Mignet N, Corvis Y. State of the art of pharmaceutical solid forms: From crystal property issues to nanocrystals formulation. ChemMedChem. 2019;**14**:8. DOI: 10.1002/cmdc.201800612

[2] Lajoinie A, Henin E, Kassai B, Terry D. Solid oral forms availabilityin children: A cost saving investigation. British Journal of Clinical Pharmacology. 2014;**78**:1080-1089. DOI: 10.1111/bcp.12442

[3] Homayun B, Lin X, Choi HJ. Challenges and recent Progress in Oral drug delivery Systems for Biopharmaceuticals. Pharmaceutics. 2019;**11**(3):129. DOI: 10.3390/pharmaceutics11030129

[4] Lee AY, Erdemir D, Myerson MS. Crystal polymorphism in chemical process development. Annual Review of Chemical and Biomolecular Engineering. 2011;**2**:259-280. DOI: 10.1146/annurev-chembioeng-061010-114224

[5] Abe H, Koyama Y, Shimono S, Kishimura H, Matsuishi K. High-pressure crystal polymorphs and multiple pathwaysin1-hexyl-3-methylimidazolium perfluorobutanesulfonate ionic liquid. Chemical Physics. 2022;**557**:111479. DOI: 10.1016/j.chemphys.2022.111479

[6] Brittain HG, editor. Polymorphism in pharmaceutical solids. In: Center for Pharmaceutical Physics. Milford. Second ed. New Jersey, USA: Informa Healthcare; 2009

[7] Li T, Mattei A. Pharmaceutical Crystals. Science and Engineering: Wiley; 2019

[8] Shi Q, Chen H, Wang Y, Xu J, Liu Z, Zhang C. Recent advances in drug polymorphs: Aspects of pharmaceutical properties and selective crystallization. International Journal of Pharmaceutics. 2022;**611**:121320. DOI: 10.1016/j.ijpharm.2021.121320

[9] Morissette SL, Soukasene S, Levinson D, Cima MJ, Almarsson O. Elucidation of crystal form diversity of the HIV protease inhibitor ritonavir by high-throughput crystallization. Proceeding of the National Academic of Science. 2003;**100**(5):2180-2184. DOI: 10.1073/pnas.0437744100

[10] Song S, Wang L, Yao C, Wang Z, Xie G, Tao X. Crystallization of sulfathiazole in gel: Polymorph selectivity and cross-nucleation. Crystal Growth & Design. 2020;**20**(1):9-16. DOI: 10.1021/acs.cgd.9b00990

[11] Katzhendler I, Azoury R, Friedman M. Crystalline properties of carbamazepine in sustained release hydrophilic matrix tablets based on hydroxypropyl methylcellulose. Journal of Controlled Release. 1998;**54**(1):69-85. DOI: 10.1016/S0168-3659(98)00002-9

[12] Cruz PC, Rocha FA, Ferreira AM. Application of selective crystallization methods to isolate the metastable polymorphs of paracetamol: A review. Organic Process Research & Development. 2019;**23**(12):2592-2607. DOI: 10.1021/acs.oprd.9b00322

[13] Zhang J, Liu Z, Wu H, Cai T. Effect of polymeric excipients on nucleation and crystal growth kinetics of amorphous fluconazole. Biomaterials Science. 2021;**9**:4308-4316. DOI: 10.1039/D1BM00104C

[14] Berry DJ, Steed JW. Pharmaceutical cocrystals, salts and multicomponent

systems; intermolecular interactions and property based design. Advanced Drug Delivery Reviews. 2017;**117**:3-24. DOI: 10.1016/j.addr.2017.03.003

[15] Kendall T, Stratford S, Patterson AR, Lunt RA, Cruickshank D, Bonnaud T, et al. An industrial perspective on co-crystals: Screening, identification and development of the less utilised solid form in drug discovery and development. Progress in Medicinal Chemistry. 2021;**60**:345-442. DOI: 10.1016/bs.pmch.2021.05.001

[16] Gao L, Zhang XR. Pharmaceutical salt hydrates of vortioxetine with maleic acid and fumaric acid: Crystal structures, characterisation and solubility performance. Journal of Molecular Structure. 2022;**1250**(2):131847. DOI: 10.1016/j.molstruc.2021.131847

[17] Gong W, Mondal PK, Ahmadi S, Wu Y, Rohani S. Cocrystals, salts, and salt-solvates of olanzapine; selection of coformers and improved solubility. International Journal of Pharmaceutics. 2021;**608**:121063. DOI: 10.1016/j.ijpharm.2021.121063

[18] Wang Y, Wang L, Zhang F, Wang N, Gao Y, Xiao Y, et al. Structure analysis and insight into hydrogen bond and van der waals interactions of etoricoxib cocrystals and cocrystal solvate. Journal of Molecular Structure. 2022;**1258**:132665. DOI: 10.1016/j.molstruc.2022.132665

[19] Zhang Y, Li Y, Chang L, Liu L, Feng Y, Wu L, et al. Improving hygroscopic stability of palmatine chloride by forming a pharmaceutical salt cocrystal of palmatine chloride-gallic acid with neutral molecule. Journal of Drug Delivery Science and Technology. 2021;**66**:102839. DOI: 10.1016/j.jddst.2021.102839

[20] Yang D, Wang H, Liu Q, Yuan P, Chen T, Zhang L, et al. Structural landscape on a series of rhein: Berberine cocrystal salt solvates: The formation, dissolution elucidation from experimental and theoretical investigations. Chinese Chemical Letters. 2021;**33**:3207-3211. DOI: 10.1016/j.cclet.2021.10.012

[21] Guerain M, Guinet Y, Correia NT, Paccou L, Danède F, Hédoux A. Polymorphism and stability of ibuprofen/nicotinamide cocrystal: The effect of the crystalline synthesis method. International Journal of Pharmaceutics. 2020;**584**:119454. DOI: 10.1016/j.ijpharm.2020.119454

[22] Braun DE, Griesser UJ. Stoichoimetric and nonstoichiometric hydrates of brucine. Crystal Growth & Design. 2016;**16**(10):6111-6121. DOI: 10.1021/acs.cgd.6b01231

[23] Clarke HDM. Crystal Engineering of Multi-Component Crystal Forms: The Opportunities and Challenges in Design. Graduates These and Dissertations. Tampa, Florida, USA: University of South Florida, Scholar Commons; 2012. Available from: https://digitalcommons.usf.edu/etd/4013/

[24] Sun Y. Improved solubility of gefitinib achieved by the water-acetone solvate. Journal of the Indian Chemical Society. 2022;**99**(1):100260. DOI: 10.1016/j.jics.2021.100260

[25] Costa RN, Reviglio AL, Siedler S, Cardoso SG, Linck YG, Monti GA, et al. New multicomponent forms of the antiretroviral Nevirapine with improved dissolution performance. Crystal Growth & Design. 2020;**20**(2):688-698. DOI: 10.1021/acs.cgd.9b01129

[26] Delori A, Friscic T, Jones W. The role of mechaochemistry and supramolecular desing in the development of pharmaceutical materials. CrystEngComm. 2012;**14**(7):2350-2361. DOI: 10.1039/C2CE06582G

[27] Erdemir D, Lee A, Myerson A. Crystal nucleation. In: Myerson A, Erdemir D, Lee A, editors. Handbook of Industrial. Crystallization Cambridge: Cambridge University Press; 2019. pp. 76-114. DOI: 10.1017/9781139026949.003

[28] Blagden N, de Matas M, Gavan PT, York P. Crystal engineering of active pharmaceutical ingredients to improve solubility and dissolution rates. Advanced Drug Delivery Reviews. 2007;**59**(7):617-630. DOI: 10.1016/j.addr.2007.05.011

[29] Aaltonen J, Allesø M, Mirza S, Koradia V, Gordon KC, Rantanen J. Solid form screening – A review. European Journal of Pharmaceutics and Biopharmaceutics. 2009;**71**:23-37. DOI: 10.1016/j.ejpb.2008.07.014

[30] Bernstein J, editor. Polymorphism in Molecular Crystals. Oxford: Clarendon Press; 2002

[31] Heinz A, Strachan CJ, Keith C, Gordon KC, Rades T. Analysis of solid-state transformations of pharmaceutical compounds using vibrational spectroscopy. Journal of Pharmacy and Pharmacology. 2009;**61**:971-988. DOI: 10.1211/jpp/61.08.0001

[32] Law D, Zhou D. Solid-state characterization and techniques. In: Qiu Y, Zhang GGZ, Mantri RV, Chen Y, Yu L, editors. Developing Solid Oral Dosage Forms: Pharmaceutical Theory and Practice. 2nd ed. Amsterdam: Academic Press; 2017. pp. 59-84. DOI: 10.1016/B978-0-12-802447-8.00003-0

[33] Nanjwade VK, Manvi FV, Shamrez AM, Nanjwade BK, Maste MM. New trends in the Co-crystallization of active pharmaceutical ingredients. Journal of Applied Pharmaceutical Science. 2011;**1**(8):1-5

[34] Chiu MH, Prenner EJ. Differential scanning calorimetry: An invaluable tool for a detailed thermodynamic characterization of macromolecules and their interactions. Journal of Pharmacy and Bioallied Sciences. 2011;**3**(1):39-60. DOI: 10.4103/0975-7406.76463

[35] Lu E, Rodríguez-Hornedo N, Suryanarayanan RA. Rapid thermal method for cocrystal screening. CrystEngComm. 2008;**10**:665-668. DOI: 10.1039/B801713C

[36] Qiao Y, Qiao R, He Y, Shi C, Liu Y, Hao H, et al. Instrumental analytical techniques for the characterization of crystals in pharmaceutics and foods. Crystal Growth & Design. 2017;**17**(11):6138-6148. DOI: 10.1021/acs.cgd.7b00759

[37] Kojnojová T, Markovičová L, Nový F. Application of thermal gravimetric analysis and comparison of polyethylene films before and after exposure in various chemical solutions. Conference Series: Materials Science and Engineering. 2021;**1178**:012029. DOI: 10.1088/1757-899X/1178/1/012029

[38] Karki S, Fabian L, Friscic T, Jones W. Powder X-ray diffraction as an emerging method to structurally characterize organic solids. Organic Letters. 2007;**9**:3133-3136. DOI: 10.1021/ol071329t

[39] Chieng N, Hubert M, Saville D, Rades T, Aaltonen J. Formation kinetics and stability of carbamazepine-nicotinamide Cocrystals prepared by mechanical activation. Crystal Growth & Design. 2009;**9**:2377-2386. DOI: /10.1021/cg801253f

[40] Bond AD. Single-crystal X-ray diffraction. In: Müllertz A, Perrie Y, Rades T, editors. Analytical Techniques in the Pharmaceutical Sciences. New

York: Springer; 2016. pp. 315-337. DOI: 10.1007/978-1-4939-4029-5_9

[41] Altomare A, Coucci C, Gatta GD, Moliterni A, Rizzi R. Methods of crystallography: Powder X-ray diffraction. European Mineralogical Union Notes in Mineralogy. 2017;**19**(2):79-138. DOI: 10.1180/EMU-notes.19.3

[42] Song Y, Cong Y, Wang B, Zhang N. Applications of Fourier transform infrared spectroscopy to pharmaceutical preparations. Expert Opinion on Drug Delivery. 2020;**17**(4):551-571. DOI: 10.1080/17425247.2020.1737671

[43] Paradowska K, Wawer I. Solid-state NMR in the analysis of drugs and naturally occurring materials. Journal of Pharmaceutical and Biomedical Analysis. 2014;**93**:27-42. DOI: 10.1016/j.jpba.2013.09.032

[44] Geppi M, Mollica G, Borsacchi S, Veracini CA. Solid-state NMR studies of pharmaceutical systems. Applied Spectroscopy Reviews. 2008;**43**(3):202-302. DOI: 10.1080/05704920801944338

[45] Carlton RA. Pharmaceutical Microscopy. New York: Springer; 2011. DOI: 10.1007/978-1-4419-8831-7

[46] Huang LF, Tong WQ. Impact of solid state properties on developability assessment of drug candidates. Advanced Drug Delivery Reviews. 2004;**56**(3):321-334. DOI: 10.1016/j.addr.2003.10.007

[47] Garnero C, Chattah AK, Longhi M. Supramolecular complexes of maltodextrin and furosemide polymorphs: A new approach for delivery systems. Carbohydrate Polymers. 2013;**94**:292-300. DOI: 10.1016/j.carbpol.2013.01.055

[48] Chierentin L, Garnero C, Chattah AK, Delvadia P, Karnes T, Longhi MR, et al. Influence of β-cyclodextrin on the properties of Norfloxacin form a. AAPS PharmSciTech. 2015;**16**(3):683-691. DOI: 10.1208/s12249-014-0259-8

[49] Garnero C, Chattah AK, Aloisio C, Fabietti L, Longhi M. Improving the stability and the pharmaceutical properties of Norfloxacin form C through binary complexes with β-Cyclodextrin. AAPS PharmSciTech. 2018;**19**(5):2255-2263. DOI: 10.1208/s12249-018-1033-0

[50] Bueno MS, Chierentin L, Bongioanni A, Nunes Salgado HR, Longhi MR, Garnero C. β-Cyclodextrin complexation as an approach to enhance the biopharmaceutical properties of Norfloxacin B hydrate. Carbohydrate Research. 2019;**485**:107818. DOI: 10.1016/j.carres.2019.107818

[51] Chattah AK, Pfund LY, Zoppi A, Longhi MR, Garnero C. Toward novel antiparasitic formulations: Complexes of Albendazole desmotropes and β-cyclodextrin. Carbohydrate Polymers. 2017;**164**:379-385. DOI: 10.1016/j.carbpol.2017.01.098

[52] Bueno MS, Miñambres GG, Bongioanni A, Chattah AK, Aiassa V, Longhi MR, et al. Exploring solid forms of oxytetracycline hydrochloride. International Journal of Pharmaceutics. 2020;**585**:119496. DOI: 10.1016/j.ijpharm.2020.119496

[53] Aguiar AJ, Krc J, Kinkel AW, Samyn JC. Effect of polymorphism on the absorption of chloramphenicol from chloramphenicol palmitate. Journal of Pharmaceutical Sciences. 1967;**56**:847-853. DOI: 10.1002/jps.2600560712

[54] Otsuka M, Matsumoto T, Kaneniwa N. Effects of the mechanical

energy of multi-tableting compression on the polymorphic transformations of chlorpropamide. Journal of Pharmacy and Pharmacology. 1989;**41**(10):665-669. DOI: 10.1111/j.2042-7158.1989.tb06337.x

[55] Ueda H, Nambu N, Nagai T. Dissolution behavior of chlorpropamide polymorphs. Chemical and Pharmaceutical Bulletin. 1984;**32**: 244-250. DOI: 10.1248/cpb.32.244

[56] Kobayashi Y, Ito S, Itai S, Yamamoto K. Physicochemical properties and bioavailability of carbamazepine polymorphs and dihydrate. International Journal of Pharmaceutics. 2000;**193**:137-146. DOI: 10.1016/s0378-5173(99)00315-4

[57] Yang W, Johnston KP, Williams RO III. Comparison of bioavailability of amorphous versus crystalline itraconazole nanoparticles via pulmonary administration in rats. European Journal of Pharmaceutics and Biopharmaceutics. 2010;**75**:33-41. DOI: 10.1016/j.ejpb.2010.01.011

[58] Healy AM, Worku ZA, Kumar D, Madi AM. Pharmaceutical solvates, hydrates and amorphous forms: A special emphasis on cocrystals. Advanced Drug Delivery Reviews. 2017;**117**:25-46. DOI: 10.1016/j.addr.2017.03.002

[59] Qiao N, Wang K, Schlindwein W, Davies A, Li M. In situ monitoring of carbamazepine–nicotinamide cocrystal intrinsic dissolution behaviour. European Journal of Pharmaceutics and Biopharmaceutics. 2013;**83**(3):415-426. DOI: 10.1016/j.ejpb.2012.10.005

[60] Jaswanth Gowda BH, Nechipadappu SK, Shankar SJ, Chavali M, Karthika P, Ahmed MG, et al. Pharmaceutical cocrystals of Efavirenz: Towards the improvement of solubility, dissolution rate and stability. Materials Today: Proceedings. 2022;**51**(1):394-402. DOI: 10.1016/j.matpr.2021.05.535

[61] Bhattacharya B, Mondal A, Soni SR, Das S, Bhunia S, Bal Raju K, et al. Multidrug salt forms of norfloxacin with non-steroidal anti-inflammatory drugs: Solubility and membrane permeability studies. CrystEngComm. 2018;**20**(41):6420-6429. DOI: 10.1039/c8ce00900g

Section 2

Crystallization and Crystal Growth

Chapter 4

Oriented Attachment Crystal Growth Dynamics of Anisotropic One-dimensional Metal/Metal Oxide Nanostructures: Mechanism, Evidence, and Challenges

Gayani Pathiraja, Sherine Obare and Hemali Rathnayake

Abstract

One-dimensional (1D) inorganic metal/metal oxide nanostructures are of significant interest due to their distinctive physical and chemical properties that are beneficial for various applications. A fundamental understanding of the guiding principles that control the anisotropy and the size of the nanostructures is essential toward developing the building blocks for the fabrication of leading-edge miniaturized devices. Oriented attachment (OA) crystal growth mechanism has been recognized as an effective mechanism for producing 1D anisotropic nanostructures. However, a limited understanding of the OA mechanism could impede the controlled fabrication of 1D nanostructures. This chapter provides a comprehensive summary on recent advances of the OA mechanism and the current state of the art on various *in-situ, ex-situ*, and theoretical investigations of OA-based crystal growth dynamics as well as the shape and size-controlled kinetics. Other competing crystal growth mechanisms, including seed-mediated growth and Ostwald ripening (OR), are also described. Further, we thoroughly discuss the knowledge gap in current OA kinetic models and the necessity of new kinetic models to elucidate the elongation growth of anisotropic nanostructures. Finally, we provide the current limitations, challenges for the understanding of crystal growth dynamics, and future perspectives to amplify the contributions for the controlled self-assembled 1D nanostructures. This chapter will lay the foundation toward designing novel complex anisotropic materials for future smart devices.

Keywords: one-dimensional nanostructures, anisotropy, oriented attachment mechanism, crystal growth dynamics, kinetic models, seed-mediated growth, Ostwald ripening

1. Introduction

1.1 Anisotropic one-dimensional nanomaterials

The controlled fabrication of crystalline anisotropic nanostructures has flourished to obtain unique functional properties and characteristics arising from their quantum confinement and nanoscale size effect for many different industrial applications. For decades, the structure-property correlations and their corresponding mechanistic principles of formation were extensively elucidated. The guiding principles and factors that affect size- and shape-control of a material, including crystal growth mechanism, phase transformation, and kinetics, have been explored for anisotropic one-dimensional (1D) nanostructures, such as nanorods, nanotubes, and nanowires [1–4]. The 1D electronic pathways to accumulate effective charge transportation and larger surface area of anisotropic nanostructures provide a profound impact in nanoelectronics and nanodevices [5]. Therefore, 1D transition metal/metal oxides crystalline nanomaterials are critical building blocks for the next-generation high-performance integrated circuits and Internet of things (IoT) applications [6–9].

Different *in-situ* interpretations have been reported to provide insights into the crystal growth of nanomaterials and how the morphology is controlled. Liquid-phase atomic force microscopy (AFM), [10, 11] cryo-transmission electron microscopy (Cryo-TEM), [12] liquid-phase TEM, [13, 14] field emission scanning electron microscope (FE-SEM), [15] and time-resolved small-angle X-ray scattering (SAXS) [12] are emergent techniques that reveal the real-space imaging of nanocrystals, intermediate structure formation, crystallization, and their growth kinetics. The efforts on in-depth investigations to understand the factors and driving forces that control the anisotropy of a nanomaterial at an atomic scale have evolved and are continuously expanding.

The significant advances of bottom-up synthetic routes that produce nanomaterials with various morphologies have been devoted to obtaining their distinctive electronic, optical, mechanical, catalytic properties. While different synthetic procedures are capable of fabricating 1D nanostructures, wet chemical methods satisfy the scalability and lower the cost. However, the development of cost-effective, greener synthesis methods that are scalable and reproducible to make stable 1D nanostructures still needs to be addressed. To accomplish this task, the foundational investigations on crystal growth mechanisms, kinetics, and phase transformation to fabricate 1D anisotropic nanomaterials are crucial. The real-time nonclassical crystallization dynamics of 1D nanostructures in solution-based synthetic processes at different temperatures and a high magnification have been limited and remain a challenge. We begin this chapter by describing the mechanisms of OA and provide insights into the OA crystal growth kinetics of anisotropic 1D nanostructures. Then we provide the recent progress of OA-based 1D metal/metal oxide nanostructures and other competing crystal growth mechanisms to lay the foundation for OA-directed anisotropic crystal growth process. We subsequently provide an insight into the characterization techniques for various *in-situ* and *ex-situ* investigation of OA-based crystal growth dynamics that makes anisotropic nanostructures along with the current state of the art. We further elaborate the theoretical studies developed along with the experimental techniques that enhance the understanding of the OA-based crystal growth mechanism and morphology evaluation. Then, we present a time-dependent crystal growth observation process of OA-based crystal growth of anisotropic copper hydroxide nanowires formed in a sol-gel colloidal system with an in-depth discussion of its growth kinetics, correlating to the sol-gel chemical kinetic reactions. Finally, we provide future perspectives of direct observation of crystal growth dynamics that enhance the fundamental understanding of nanoscale colloidal assembly mechanisms to achieve morphology-controlled properties of nanomaterials for future needs in advanced applications.

1.2 Oriented attachment (OA) mechanism

In recent years, nonclassical oriented attachment (OA) mechanism has received a great attention to produce various morphologies including zero-, one-, two-, and three-dimensional nanostructures with their controlled structural properties [16]. More importantly, it is an effective mechanism for anisotropic crystal growth of nanostructures, inclusion of defects to the crystal structure, and the formation of branched nanostructures and highly ordered mesocrystal structures [17]. This key crystal growth mechanism can be found in natural biomineralization processes that show unique physicochemical properties [18]. In 1998, the observation of anisotropic chain of TiO_2 anatase nanocrystals attachment across the {112} crystal facets using a high-resolution transmission electron microscope (HR-TEM) by Penn and Banfield had contributed to this discovery of the OA mechanism [19]. The thermodynamic reduction in the surface energy of TiO_2 crystal facets during the attachment process has driven the structural anisotropy [19, 20].

1.2.1 Concept and mechanism of OA

OA is the self-assembly of adjacent nanocrystals for a specific crystal facet attachment to form a secondary single nanocrystal through the Brownian motion [19, 21]. While Penn and Banfield visually observed the OA-based crystal growth using TEM, Moldovan and coworkers proposed the detailed mechanistic principles in 2002 using both analytical and molecular-dynamics (MD) simulation methods [22, 23]. According to their model named as "Grain-Rotation-Induced Grain Coalescence (GRIGC) mechanism," the adjacent primary nanocrystal colloids in a solution freely rotate to match the perfect coherent grain-grain interface. After they met perfect crystal facet match, these nanocrystals start coalescence to eliminate the common grain boundaries and form a single large nanocrystal [22]. This is the thermodynamic reduction of the crystal surface energy of nearby nanocrystals that drive to minimize the high energy surfaces. **Figure 1** shows the schematic representation of GRIGC mechanism to form an OA-based 1D nanostructures such as nanorods and nanowires.

1.2.2 Characteristic of OA

The anisotropic nanostructures produced using OA process provide unique characteristics in size, crystal structure, and kinetics. The constant diameter during the growth is one of the unique advantages of OA mechanism by attaching of nanocrystals to the tip of the growing 1D nanostructure [4]. This is very similar to the polymerization reaction processes. Also, there is a capability to predetermine the diameter of the nanostructure, by observing the diameter of the monodispersed nanocrystals. If the diameter of 1D nanostructure is below 10 nm, it offers interesting and improved characteristics. For example, ultrathin nanowires provide new surface determined structures, with tunable surface chemistries, higher surface area, and colloidal stability for different applications [24–27]. Nanocrystal that grows following the OA mechanism has abundant defects such as twin planes, stacking faults, and misorientations

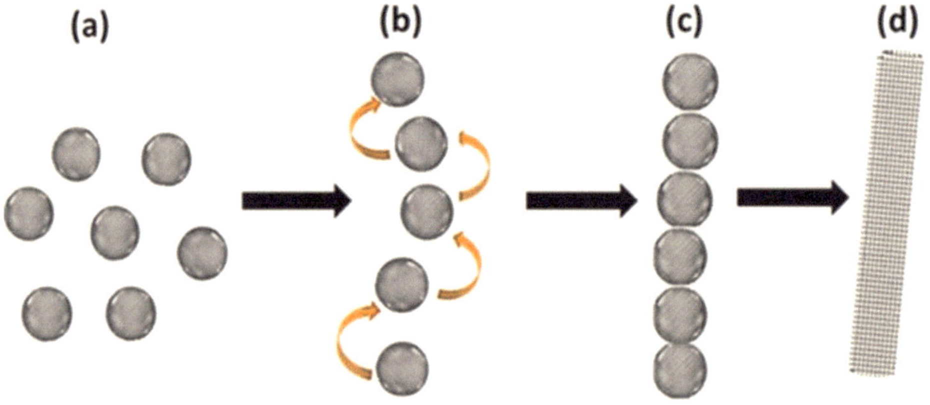

Figure 1.
Schematic representation of stages on OA crystal growth mechanism to form an anisotropic nanostructure. (a) Primary nanocrystal colloids in a solution, (b) Rotation of nanocrystals to match the crystal facets of nanocrystals, (c) Coalescence of nanocrystals along a specific crystallographic orientation, and (d) Formation of a single crystal 1D nanostructure.

[28, 29]. This defect formation occurs due to the crystal lattice mismatch during the nanocrystal attachment process.

OA-based chemical reactions follow second-order kinetics, and there are three kinetic models developed to explain the 1D crystal growth kinetics of nanocrystals based on collision of nanocrystal numbers [2]. These existing kinetic models were developed based on the diameter growth of the nanocrystals. The crystal growth rate is correlated exponentially with the nanocrystals' surface energies in OA growth [30]. The crystal growth rate is higher in high surface energy planes; therefore, the final crystal facet of the product is the lower than the surface energy crystal plane. The morphology or the size of OA-based 1D nanostructures can be modulated by preferentially adsorbing solvents, ligands, and surfactants. They adsorb selectively with different nanocrystal binding affinities onto specific crystal facets [31]. Studies of shape-controlled synthesis of different 1D nanostructures using selective adsorption of surfactants have been reported [32, 33].

1.2.3 Recent progress of OA-based metal/metal oxide nanostructures

To date, OA mechanism has been used to fabricate different metal/metal oxides/ metal hydroxides 1D nanostructures to use in different applications. Few recent examples of controlled synthesis of different nanorods and nanowires are Au [34], NiO [35], Sb_2O_3 [36], Co_3O_4 [35], ZnO [37–39], TiO_2 [40], MnO_2 [41],CuO [42, 43], Cu $(OH)_2$ [44], and GaOOH [45] for optoelectronic devices, electrochemical devices, and supercapacitors. Although these metal/metal oxide nanostructures have been cited as OA mechanism-based nanostructures, only few studies have investigated the crystal growth mechanism with the understanding of their guiding principles. The very first studies of Penn and Banfield investigated imperfect attachment of anatase TiO_2 nanocrystals by describing their driving tools of nanocrystals rotation and coalescence observed by HRTEM [19]. Recently, Zhang and coworkers demonstrated direction specific van der Waals attractions between two rutile TiO_2 nanocrystals [46]. They utilized AFM probe technology with environmental TEM (ETEM) to elucidate the relationship between the orientation, contact area, and surface roughness of nanocrystals. Another outstanding investigation of OA process is the direct observation of iron oxyhydroxide nanoparticles by liquid cell HR-TEM [47]. This study reveals the importance of electrostatic interactions for the lattice match attachment of nanocrystals, and they have successfully attempted to measure the translational and rotational accelerations. Leite *et al* performed an *ex-situ* observation of the OA crystal growth process of SnO_2 nanocrystals at room temperature using HR-TEM [48]. Another example of investigating the driving forces of OA process is ZnO nanocrystals using the HR- TEM coupled with x-ray diffraction (XRD) in the gas phase in the presence of a constant electric field [49]. They suggested that the OA process was dominant due to the increased dipole interactions of ZnO nanocrystals with the electric field. The recent investigation of OA-based Au nanocrystals using liquid cell-TEM provided evidence of the control of crystal facets by capping ligands to adsorb on different surfaces at an atomic scale [14]. The real-time observation of nanobridge formation between adjacent Au nanocrystals and then rearrangement of single nanocrystal via grain boundary migration using liquid cell-TEM is a remarkable investigation of OA process, which corresponds to the self-assembly of nanocrystals via "jump-to-contact" mechanism [50].

1.3 Other crystal growth mechanisms

1.3.1 Seed-mediated growth

Seed-mediated growth is a common growth mechanism to produce well-controlled crystalline noble metal nanostructures. It involves two main steps: nucleation and growth, as shown in **Figure 2**. In nucleation, the metal precursors undergo the reduction process to form zerovalent metal atoms that self-assemble into small clusters to further grow into stable nuclei. These crystalline nuclei act as monodispersed seeds for the subsequent growth of metal nanostructures during the growth stage [51]. This mechanism can be divided into two main types based on their temporal and spatial differences of the nucleation and growth stages, which includes homogeneous and heterogeneous nucleation [52]. In homogeneous nucleation, the seed nanocrystals are generated and followed by nucleation and growth processes in the same chemical reaction. In contrast, the seed nanocrystals are synthesized separately and then added into a growth solution to further growth of nanocrystals in the heterogeneous nucleation.

The surfactants/capping agents such as cetyltrimethylammonium bromide (CTAB) and cetylpyridinium chloride (CPC) can facilitate the controlled crystal facet directed growth of nanostructures via preferential adsorption. Although this is a very

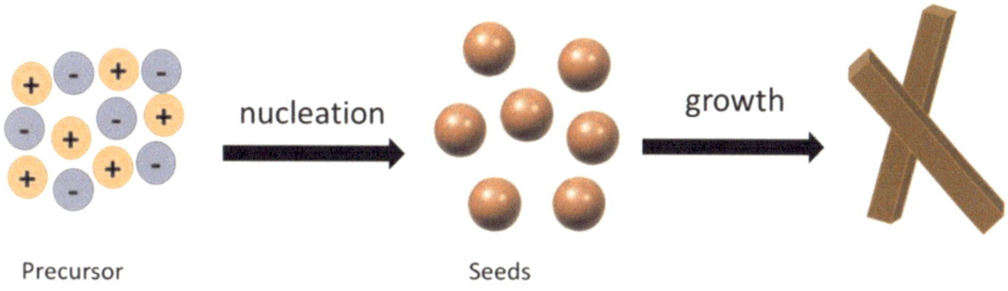

Figure 2.
Schematic representation of seed-mediated growth to form an anisotropic nanostructure.

versatile process, the major drawback of this mechanism is the limited length growth of nanostructures due to the weak interaction forces [9]. Early attempts to produce Au nanorods designed a three-step seeding protocol by Murphy and Obare team [53]. **Figure 3** shows the different aspect ratio of synthesized Au nanorods at different pH values in the solution. This modified seed-mediated method produced longer nanorods at higher pH value with the high yield of product.

1.3.2 Ostwald ripening (OR) mechanism

The classical crystal growth mechanism is the Ostwald Ripening (OR) that describes the smaller crystals grow into the larger crystals through the diffusion to reduce the total surface free energy (**Figure 4**) [54–56]. It is a concentration gradient-induced process around the surface of particles that follows the Gibbs-Thompson relation, as shown in Eq. (1) [56, 57]. Therefore, it results in spherical particles most often, which are in the micrometer size diameter range. However, this mechanism is most often unable to describe the crystal growth in the nanoscale. Further, the particles with similar crystal symmetries can also obtained due to the similar crystal facet surface energies [56].

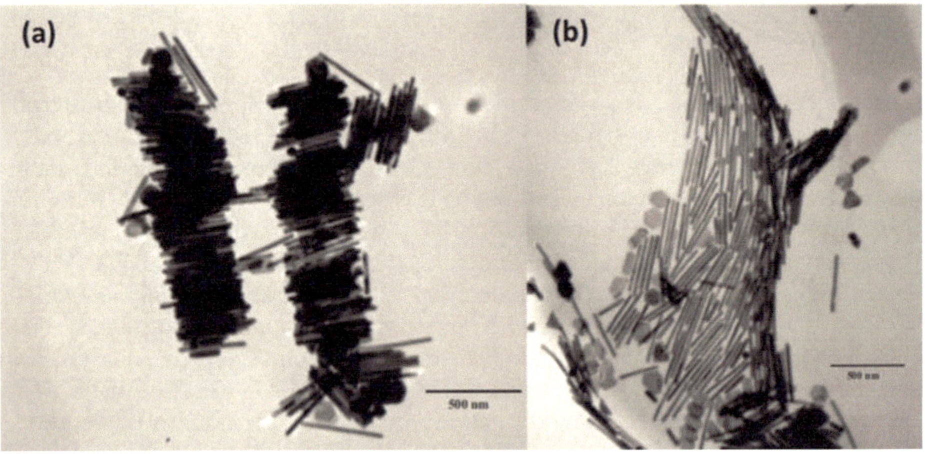

Figure 3.
TEM images of seed-mediated growth-assisted Au nanorods after one round of purification having (a) aspect ratio ~18 at pH = 3.5, and (b) aspect ratio ~25 at pH = 5.6. These images were reproduced from Brantley et al. *[53] with permission from Willey Online Library.*

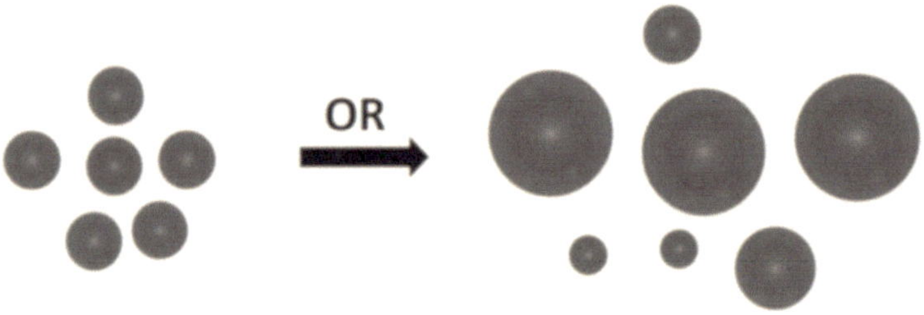

Figure 4.
Schematic representation of OR-based crystal growth.

$$C_r = C_e\, exp\left[\frac{2\gamma\Omega}{RTr}\right] \tag{1}$$

where C_r is the equilibrium concentration at the surface of the particle, C_e is the equilibrium concentration at a plane interface, γ is the surface free energy, R is the universal gas constant, Ω is the molar volume of the particle, T is the temperature, and r is the radius of the particle.

Since there is no involvement of crystal facets between the neighboring crystals, in the OR process the crystal defects are less than OA based structures. Although OA and OR the growth processes involve the growth of nanocrystals, the size gradually increases in OR mechanism, while particle size increment is stepwise in OA. Most often, both OR and OA mechanisms involved during the many synthesis processes and determine the final morphology.

1.4 OA and OR kinetic models

The kinetics of crystal growth of nanostructures are dependent on the nature of the material, the type of crystal facet interface, the solvent, temperature, and concentration of the surfactant [28]. Both OR and OA crystal growth mechanisms can occur simultaneously, and their kinetic models were developed with respect to the diameter growth of a nanostructure. However, it is possible to change the crystal growth into an OA mechanism by hindering OR mechanism initially by introducing surfactants [28, 58]. It is very important to know the characteristics of the OR crystal growth mechanism to distinguish from the OA mechanism. As the OR mechanism is a diffusion-controlled process, it is favored thermodynamically with saturated solution that dissolves nanoparticles. In contrast, OA mechanism dominates in undersaturated conditions [59]. The crystal growth kinetic models of these OR and OA mechanisms also behave differently. OR mechanism-based kinetics models follow first-order kinetics while OA-based kinetic models follow second-order kinetics [60, 61].

The OR kinetic model was proposed by Lifshitz, Slyozov, and Wagner named as LSW kinetic model [2, 62]. The first-order equation for the linear crystal growth rate can be expressed by an exponential function as follows [57].

$$\mathrm{D^n - D_0^n = k(t - t_0)} \tag{2}$$

where D and D_0 are the mean particle sizes at time t and t_0, k is a temperature-dependent rate constant, n is an exponent related to the coarsening mechanism through the diffusion.

The crystal growth kinetic process in OA is complicated than in OR kinetics since nanocrystals in the different stages of the reaction go through the collision and coalescence during the attachment. Therefore, three kinetic models have developed to explain the OA-based crystal growth of nanocrystals [2, 61]. They are: (1) primary particle-primary particle model (A_1+A_1), (2) primary particle-multilevel particle (A_1+A_i) model, and (3) multilevel particle-multilevel particle (A_i+A_j) model, developed by Smoluchowski based on the collision between number of particles [2]. **Table 1** demonstrates the population growth matrixes of these three OA kinetic models. However, these modified Smoluchowski equations can describe the nanoparticle's diameter growth of the reaction. The kinetic models that describe the elongational growth of 1D nanostructures are essential to explain the controlled

Oriented-attachment-based kinetic models		
A_1+A_1 model	**A_1+A_i model**	**A_i+A_j multistep model**
$d = \frac{d_0\left(\sqrt[3]{2}k_1t+1\right)}{(k_1t+1)}$	$d_{eq} = \sum N_k d_k^4 / \sum N_k d_k^3$	$d_{eq} = \sum N_k d_k^4 / \sum N_k d_k^3$
d and d_0 are mean diameter at time t and t_0	d_{eq}—average particle size	d_{eq}—average particle size
	d_k—size of the secondary particle containing (1+i) primary particles	d_k—size of the secondary particle containing k primary particles
k_1 = rate constant	N_k—number of secondary particles	N_k—number of secondary particles
	Collision Matrix	k × k matrix
	$[(1+2)(1+3)(1+4)(1+5). \ldots (1+(i-1))\ \ (1+i)]$	$\begin{bmatrix} (1+1)(1+2)(1+3)(1+4)\ldots(1+k) \\ (2+2)(2+3)(2+4)\ldots(2+k) \\ (3+3)(3+4)\ldots(3+k) \\ (i+i)\ldots(i+k) \\ ((k-1)+(k-1))((k-1)+k) \\ (k+k) \end{bmatrix}$

Table 1.
Three basic kinetic models OA-based crystal growth of nanoparticles [2].

fabrication of anisotropic nanostructures. Very few metal oxide/hydroxides anisotropic structures were explained by fitting the existing three kinetic models to understand the OA crystal growth mechanism, and they are not satisfactory enough to understand the growth rates.

2. Characterization techniques to visualize OA crystal growth dynamics

An in-depth understanding of the guiding principles of OA mechanism that dictates the attachment of adjacent nanocrystals toward a specific crystallographic orientation is still critical for the progress of miniaturization and high aspect ratio of anisotropic 1D nanostructures. The visualization tools including *in-situ* and *ex-situ* electron microscopy techniques and computational simulation methods have been investigated to reveal the OA crystal growth processes. It was found out multiple factors influence the OA process such as the type of the solvent or surfactant in the medium and their polarity, size and shape of the primary nanocrystal, temperature, and concentration of precursors [28, 58, 63]. The size and shape can be controlled by changing either surface energies of the crystal facets and the external growth environment [31]. However, the deeper understanding to obtain ultrathin 1D nanostructures is still in its infant stage. In this section, we discuss the main characterization tools that visualize the OA crystal growth process performed in different anisotropic nanomaterials and their findings.

2.1 Ex-situ investigation techniques

The *ex-situ* investigation is the technique that performs outside of the reaction process such as after the drying or processing. The disadvantage of this method is that artifacts induced to the system after the post process may affect the interpretation of the analysis. Therefore, it needs the careful analysis with other supporting techniques. Transmission electron microscopy (TEM) is an advanced, versatile, and standard *ex-situ* tool to obtain the structural and chemical information of the nanocrystalline materials such as elemental composition and mapping, size, shape, and crystallinity [64]. Selective area electron diffraction (SAED) feature associated with the TEM is used to determine the crystal structure and their crystallographic orientation in a specific area of the nanostructure [65]. This high accelerated electron beam related technique is enable to characterize the uniformity of nanomaterials [64]. The scanning transmission electron microscope (STEM) mode is one of the recent advancements of TEM, which generates images by performing a raster scan on the surfaces of nanostructures [64]. This STEM detector is coupled with a high-angle annular dark-field (HAADF) detector and the energy-dispersive X-ray detector (EDX or EDS) that provides the elemental composition and mapping in the nanostructures [66].

Today's HR-TEM can provide atomic-resolution intrinsic structure, crystal lattice information combined with the chemical information of a single nanocrystal [67]. Therefore, it is one of the frontier-characterization tools that provide the understanding of the size and shape-controlled anisotropic nanostructures. One of the extensive studies of identifying the possible crystal growth mechanism of Ag nanowires was performed by Murph and her team utilizing both HR-TEM analysis and molecular dynamics (MD) simulations [9]. They have monitored the intermediate stages of the synthesis to visualize the coarsening process of similar crystallographic facets of neighboring nanocrystals. MD simulation results have suggested that the dipole-dipole attraction causes the preferential crystallographic attachment of hydroxide ions on the surfaces of nanocrystals to produce these ultra-long Ag nanowires.

HR-TEM provides crystal defect information such as twins, misorientation, tacking faults, and phase transformation, and it is an important visualization technique to identify OA crystal attachment process [64, 68]. The very early reports of Penn and Banfield demonstrated the crystal defects of TiO_2 nanocrystals including the edge, screw, and mixed dislocations by looking at the lattice fringe details of crystals using HR-TEM [19]. They referred such defects as imperfect oriented attachment, which can be expected in natural and experimental conditions.

Time-dependent XRD technique is another important characterization tool used to investigate crystal growth planes during the OA growth process. Although it is not a standalone technique to identify a crystal growth mechanism, it provides a platform to track dynamics of crystal planes growth at different stages of OA process combined with the selective area electron diffraction (SAED) of TEM [69]. Our group reported such valuable investigation to observe the gradual coalescence, reorientation of crystal facets during the OA-based sol-gel-derived process of ultrathin $Cu(OH)_2$ nanowires [44]. **Figure 5** shows the time-dependent SAED patterns and powder XRD traces of $Cu(OH)_2$ nanoarrays and nanowires at the different stirring time intervals followed by aging in a base-catalyzed sol-gel chemical process.

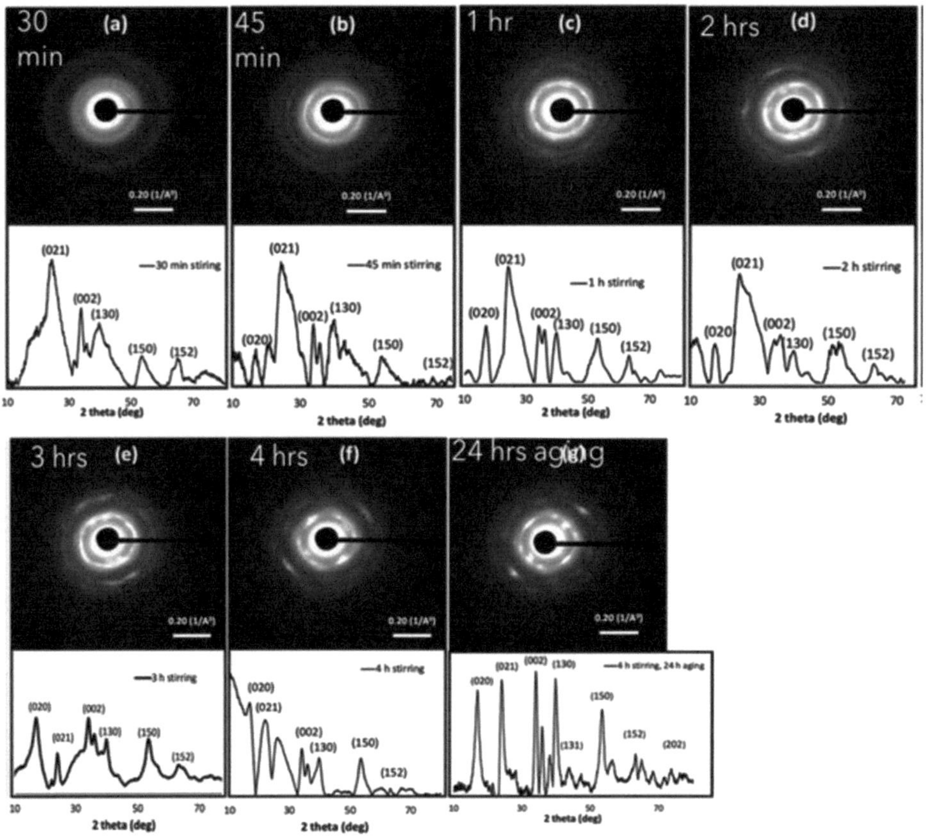

Figure 5.
Time-dependent SAED patterns along with powder XRD traces of $Cu(OH)_2$ nanocrystals at different stirring and aging time. Reproduced from reference [44] with permission from the Royal Society of Chemistry.

2.2 In-situ investigation techniques

The *in-situ* investigation tools are prominent techniques to accurately investigate the crystal growth mechanisms as it can perform real-time monitoring in the reaction solution without any further modifications. The recent advances of liquid-phase TEM is one of the leading tools for the direct visualization of different nanostructures [47, 70–73]. The high spatial and temporal resolution of the liquid-phase TEM allows to comprehensively understand the underlying growth mechanism at an atomic scale. It also provides the information about crystals orientation and crystal defect formation during the nanocrystal's attachment [74]. As a result, our understanding of nanocrystal nucleation, growth, and their dynamics has accelerated. However, it requires a careful interpretation as it has the electron beam effect, substrate effect, and some synthesis procedures are complex and incompatible with liquid phase-TEM [75, 76]. Therefore, this tool has been limited to few synthetic process and materials although it facilitates real- time monitoring. However, these *in-situ* techniques alone cannot be utilized to confirm the growth mechanisms. They should undergo the *ex-situ* analysis as a supporting information to prove the direct observation analysis.

The direct observation of metal hydroxide/oxide growth is less as most common metal oxide synthesis methods are not compatible to observe under liquid-phase TEM. The important calculations of translational and angular speeds of iron oxyhydroxide nanoparticles were performed during the OA growth by Li and the team for the first time using liquid-cell TEM [47]. Furthermore, the Pb_3O_4 nanocrystals coalescence, and growth rates were determined by another group during the OA growth along the [002] crystal facet using liquid-cell TEM [77]. Another recent study demonstrated the fivefold twinned Au crystal domains formation using real-time HR-TEM imaging [78]. *In-situ* monitoring ZnO nanorod formation reached a milestone by providing new insights into the dynamics of OA [79]. The driving forces and torques for both aggregation and alignment were determined using the individual trajectories and attachment events of several ZnO nanoparticle pairs. The OA mechanism was confirmed using lattice fringes observations and its Fast Fourier transform (FFT) analyses. Investigation on oriented attachment of Au nanoparticles was performed by Zhu *et al* using direct observation of liquid TEM. In this ligand-controlled reaction, they observed that the overlapped ligands follow the rotation into {111} orientation. The calculated ligand binding energy on {111} crystal facets is lower than that of other crystal facets, which causes the preferential attachment of Au nanoparticles [14].

Small-angle X-ray scattering (SAXS) is another important *in-situ* tool to get qualitative and quantitative understanding of the OA mechanism at different stages of the process. It can accurately determine the shape and size of nanostructures [80]. Recently, the nucleation and growth of Au nanoparticles were successfully investigated to understand the mechanism and kinetics using SAXS technique in the solutions [81–83]. An early attempt of direct probing of TiO_2 nanorods in the reaction solution was performed by Tsao and the group at different temperatures for different reaction times [84]. The time-dependent temporal evolution of SAXS profiles revealed the spontaneous alignment of the quasi-spherical particles with the time from initially formed spherical particles. They confirmed that this observation combines with the HR-TEM analysis.

2.3 Computational simulations

The computational simulation techniques are a growing field to interpret experimental observations to reveal the detailed information of key growth mechanisms.

The main simulation methods are molecular dynamics (MD), density functional theory (DFT), ab initio, and Monte Carlo (MC) calculations. These methods provide insight into the thermodynamics, kinetics, and driving forces of crystal growth mechanisms [85, 86]. Although computational simulations involved with faster timescales than experiments, they provide a valid approach to understand crystal growth dynamics with a good agreement with experimental results.

The current arguments of the driving force for OA mechanism are controversial. The main assumption of OA is the reduction of surface energy of nanocrystals that causes the thermodynamic driving factor for the spontaneous attachment of nearby nanocrystals [19, 87]. However, recently some research works have suggested that OA is driven by physical driving forces such as van der Waals interactions [88] and/or dipole-dipole interactions [89, 90]. Computational simulations are important to compare these different driving factors and then finally find out the primary contributing factor in different systems. Zhang and Banfield successfully analyzed these key factors separately by treating the system differently [86]. They demonstrated that Coulombic interactions are predominant when nanoparticles are close to each other in a solution, while van der Waals interactions are dominant when nanoparticles are far apart in a solution.

The main fortuitous advantage of combination of atomistic computer simulation techniques with experimental characterization methods is to validate the assumptions of crystal growth mechanism and controlling factors observed in the experiments. MD simulation is useful to explain the driving forces for the *ex-situ* TEM observations of different nanostructures. Zhang *et al* described the direction-specific interaction forces that can create torque to align adjacent ZnO nanocrystals and induce OA to form ZnO nanorods using MD simulations [91]. Very early studies of MD estimated the free energy changes of MgO nanoparticles due to the OA aggregation toward specific orientation in vacuum [92]. Another study used ab initio methods to calculate the surface energies of crystal faces of Sb-doped SnO_2 nanocrystals to predict the final morphology, which is observed from *ex-situ* TEM [93]. Murph *et al* also used MD studies to validate their HR-TEM experimental data on Ag nanowires and describe the mechanism to form penta-twinned nanowires growth along the [110] facet [9].

Very recent work demonstrated the solvent effect for the probability of perfect and imperfect OA mechanism by mapping the crystal growth dynamics of Ag nanocrystals in contrast to the solid state without the solvent in the medium [94]. This study is useful to develop new solvent directed strategies to control the specific crystallization processes to obtain the desired final product. Furthermore, Sayle and the group predicted the mechanical properties of CeO_2 nanorods and nanochains using atomistic computer simulation with the in detailed discussion of the effect of dislocations and grain boundaries for mechanical properties [95].

3. Novel approaches to understand OA longitudinal growth dynamics and kinetics of anisotropic 1D nanostructures

One-dimensional transition metal/metal hydroxides/oxide nanostructures are ideal building blocks for the miniaturization of devices. Understanding the mechanism and kinetics that control the final morphology of nanostructures is essential for the progress toward materials design for desired applications. Early studies show that surfactants/organics are one of the guiding factors for the OA growth. However, the presence of organics is not compulsory to drive the OA mechanism [96]. There are enough examples to demonstrate OA-based nanostructure formation without any organic additives [37, 97, 98]. The temperature and pH value of the reaction are other driving forces for OA mechanism [99]. However, the attempts observing the mechanism to elucidate the kinetics of OA-based elongation of 1D nanostructures are still under studied. Few reports that attempted to explain the crystal growth kinetics of one-dimensional nanostructures along certain specific direction are described in this section.

Gunning and the team presented the multistep kinetic model to explain the elongation of CdS nanorods using dipole-dipole interactions in the presence of amine [100]. They suggested that the amine concentration drives the end-to-end attachment of nanocrystals by minimizing the surface energy to align the nanocrystals. Another study investigated the roles of van der Waals attractions and Coulomb interactions for OA growth of nanorods and nanowires [101]. These 1D nanostructures are assumed to be cylindrical shape and nanocrystals to be spherical shape. The Hamaker's particle-particle model was used to calculate van der Waals attractions in the system to explore the effect of head-to-head attachment of nanorods. It was found that the role of van der Waals interaction was to guide the 1D nanostructures formation. Furthermore, they calculated the role of Coulomb interactions for OA growth using Coulomb's law. They have considered the nanoparticle-nanorod separation to find out the effect of different parameters on Coulomb-interactions-based OA growth. He and coworkers also investigated the parameters associated with van der Waals interactions to drive the OA growth for nanorod formation [102].

Our group achieved a significant milestone by developing novel chemical kinetic models for the first time to describe the OA-directed crystal growth kinetics in the sol-gel chemical process [103]. By assuming that the sol-gel process is a quasi-homogeneous system, dimensional changes in the nanocrystals at three stages of the sol-gel process that forms $Cu(OH)_2$ nanowires were monitored using *ex-situ* TEM. It was found out that nanowire growth follows second-order, zeroth-order, and zeroth-order sigmoidal Boltzmann models during the hydrolysis and condensation, first stage of polycondensation, and the second stage of polycondensation phases, respectively, as shown in **Figure 6**. The sigmoidal growth curve represents three characteristic phases including the initial lag phase, growth phase, and saturation phase that correspond to the three stages of the sol-gel process. The first stage of hydrolysis and

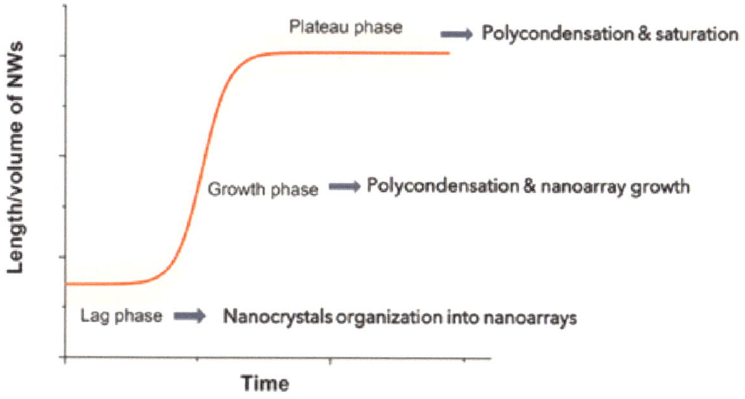

Figure 6.
The sigmoidal plot that describes three stages in the sol-gel process to fabricate ultrathin $Cu(OH)_2$ nanowires as described in reference [103].

condensation process follows second-order kinetics to the self-organization of nanocrystals to form nanochains. Then the directional elongation of nanoarrays along a specific crystal facet occurs during the first stage of polycondensation following zeroth-order kinetics. Finally, the second phase of polycondensation demonstrated a steady saturated growth that forms $Cu(OH)_2$ nanowires following zeroth-order kinetics. To validate these kinetic models, we utilized statistically significant datasets, higher the regression coefficient (R^2) values, which is at 95% levels of confidence for three replicates.

We identified these important stages of OA mechanism, including nanocrystal formation and their self-assembly toward orientation, nanoarrays, and nanowires formation by observing time dependent *ex-situ* TEM images. **Figure 7** demonstrates the formation of primary $Cu(OH)_2$ nanocrystals (sols) during the hydrolysis and condensation. We successfully captured the neck initiation between two nanoparticles, which shows same crystallographic orientation. Similar observations were reported by Wang *et al* for the PbSe nanocrystals using *in-situ* liquid cell TEM [104]. The crystallographic lattice orientation and the epitaxial growth nanocrystals were determined by the (Fast Fourier Transform) FFT and SAED, respectively. The d-spacing of a nanocrystal confirms the $Cu(OH)_2$ crystal structure.

The HR-TEM image in **Figure 8a** shows the alignment of nanocrystals forming nanocrystal chain after 5 min of stirring. The coalescence of nanocrystals is incomplete, and lattice orientations rotation is visible toward the same crystal facet orientation. Therefore, our study further confirms the evidence of continuous fusion and rotation of different crystal-faceted nanocrystals to match their lattice orientation in the OA mechanism. The next image in **Figure 8b** shows the progress of nanocrystals orientation with smoothing the edges after 1 hour stirring time. The formation of nanowire after 4 hours stirring is shown in **Figure 8c**, and we can observe neck elimination between nanocrystals and a uniform distribution of lattice orientation in this nanowire. The inset FFT pattern confirms the atomically coherent single crystal, which is the final product of the OA mechanism. This study provides insight to produce anisotropic

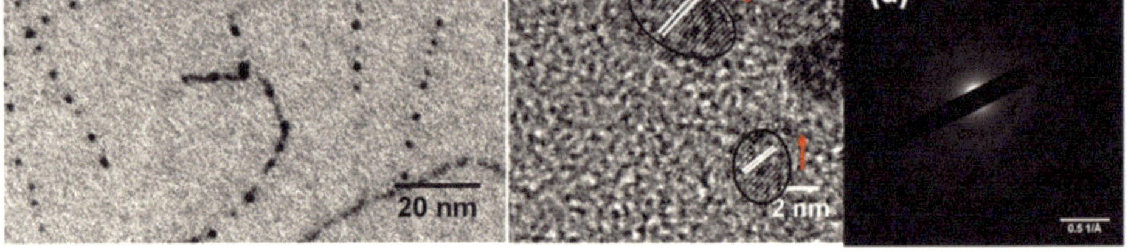

Figure 7.
(a) TEM image of nanocrystal seed formation during the addition of base, (b) HR-TEM image of $Cu(OH)_2$ *nanocrystals with respective lattice d-spacing for [022] facet; arrow marks indicate crystal lattice orientation of different nanocrystals, (c) respective FFT image, and (d) its SAED pattern. Reproduced with permission from reference [103] Copyright {2022} American Chemical Society.*

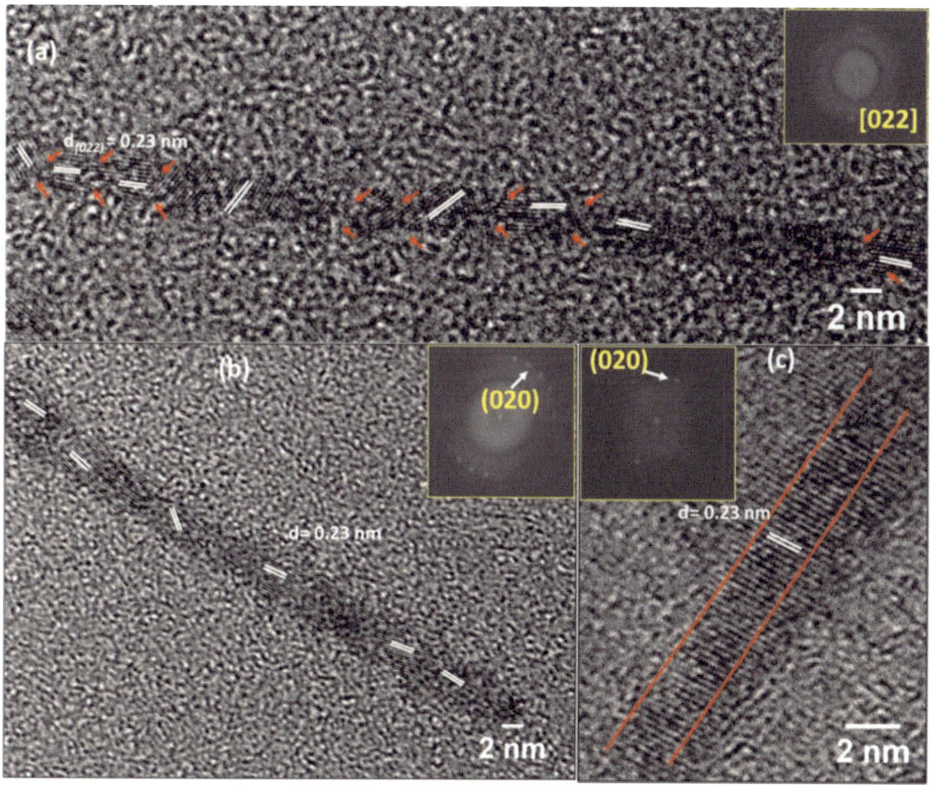

Figure 8.
HR-TEM images of nanoarrays at: (a) 5 min, (b) 1 hr, (c) 4 h stirring time, respectively. (Inset is its FFT, and the red arrows indicate necks between nanocrystals) Reproduced with permission from reference [103] Copyright {2022} American Chemical Society.

nanostructures of other metal oxides via the sol-gel processes by tailoring the reaction parameters, such as reaction time, temperature, solvent, and the pH.

4. Summary, challenges, and future prospective

Inorganic one-dimensional metal/metal oxide nanostructures play important roles in various miniaturized electrical, optical, and energy devices. The exploration of nanostructure-property relationship in terms of size, shape, interaction, and crystal growth of nanostructures is essential for the rational design and synthesis of tailored anisotropic nanostructures. However, the reaction kinetics and crystal growth mechanisms are not well understood, and the further development of both theories and experiments is expected. The OA mechanism, which is the effective mechanism to fabricate anisotropic structures, requires understanding their guiding factors that enable the rational design of different 1D nanomaterials. The driving forces for the preferred attachment orientations, origin of the force field between nanocrystals, collision trajectories, surface energies of crystal facets are needed further investigations, using direct *in-situ* observations, *ex-situ* methods, and computer simulations. Although there is a development of experimental and theoretical investigations of OA mechanism for 1D nanostructure fabrication, certain areas still need to be addressed for the development of next-generation high-performance devices.

4.1 Limitations of *in-situ* experimental techniques to investigate guiding factors at an atomic scale

The recent advances in the *in-situ* experimental techniques facilitate the direct real-time observations of the OA crystal growth. The different key factors may be synergistic in driving OA growth successfully in a complex environment. Therefore, identifying individual contributions and their significance is challenging. *In-situ* techniques allow us to track the trajectory of nanostructures such as nucleation, nanocrystals formation, self-assembly, coalescence and growth, and the physical behavior in a solution. These facilities are not available and limited in *ex-situ* experiments.

The development of liquid cell TEM has led to make it capable to visualize at an atomic scale and has contributed significantly for the reasonable growth in this field. However, there are limitations and challenges during *in-situ* imaging of nanomaterials. The electron beam irradiation effect, localized heating effects that create artifacts or defects in the sample, substrate effects are common problems in liquid cell TEM. Currently, these effects can be mitigated to some extent by controlling the beam exposure, flux, and current, to reduce the electron beam effects and radiation damage of nanostructures and introduction of different sample stages to promote heat dissipation during imaging. Some other potential solutions are introduction of high detection efficiency camera systems and couple compressive sensing software system into the TEM. However, there are still more unstable materials yet to be explored by *in-situ* imaging as they are more sensitive to the electron beam. It is required to optimize the parameters in the TEM to reduce the beam exposure and damages with different materials. Although this is very time-consuming and complicated technique, it is worth to generate information using different nanomaterials that enhances our fundamental understanding on OA mechanism. Another challenge of *in-situ* imaging is having complicated synthesis procedures to fabricate metal/metal oxide 1D nanostructures. The new facile and efficient solution route-based synthesis procedures need to be in place as they are more convenient to investigate fundamental principles using *in-situ* experiments. The rapid progress and introduction of new features to TEM or other instruments are also advancing these investigations.

4.2 The invalid existing OA kinetic models to describe the elongational crystal growth of anisotropic structures

The existing modified Smoluchowski kinetic models can only describe the diameter growth in the OA mechanism. The use of these kinetic models for different nanocrystals is very limited and understudied. The kinetic studies of OA-directed crystal growth processes are helpful to address the development of different synthesis methods of size and shape-controlled 1D nanostructures. However, there is less development of kinetic models to describe the elongation growth of 1D nanostructures. Although few studies are attempted to describe the OA elongation kinetics, they are limited to their specific system and haven't modeled using experimental observations. The theoretical kinetic models are essential to develop, introduce, and validate with multiple experimental observations. It could further expand to describe the OA crystal growth processes in the presence of surfactants or organic additives.

4.3 Lack of efficient and greener synthesis methods to make ultrathin anisotropic nanostructures

The controlled synthesis of ultrathin anisotropic nanostructures is limited to few materials, and the more efficient, scalable, greener, facile, and reproducible synthesis techniques are rarely introduced. The attempts of developing new strategies are needed to enhance the prospects for the preparation of controlled 1D nanostructures. Although there is a rapid progress on OA-based 1D metal/metal oxide nanostructures without any organic additives, we are still far away from the full understanding of OA crystal growth. The new efficient and greener synthesis methods will certainly open doorways toward rationally designing different ultrathin nanostructures. Although most common method to produce 1D noble metal nanostructures is seed-mediated growth methods, they are ineffective methods to make long nanostructures such as nanowires. Therefore, enough attempts are required to change the aspect ratio of 1D noble metal nanostructures with in-depth understanding of the controlling factors. The new solution-based wet chemical strategies are required to develop unrestrictedly size and shape-controlled 1D nanostructures to get the advantages of cost-effectiveness, energy effectiveness, and scalability of the process.

4.4 The insufficient theoretical investigation to improve fundamental understanding

The theoretical investigations further advance our fundamental understanding on OA crystal dynamic growth processes. Molecular dynamics, DFT simulations, molecular mechanics, ab initio methods are a few tools that can be used to develop theories and strengthen the understanding of the OA mechanism. The necessity of these theoretical tools combined with experimental techniques is increasing demand for further insight into the OA growth at different levels including molecular, atomic, and crystal lattice scale. If we combine real-time crystal growth dynamics with simulations of interparticle forces and interactions, it will provide the explanation of guiding factors quantitatively. However, the utilization of theoretical models to develop more accurate experimental OA growth models is limited, thus necessitating collaborative experimental work for further advancement.

5. Conclusions

In conclusion, the in-depth discussion of potential of "oriented attachment" mechanism has been overviewed in size and shape-controlled inorganic metal/metal oxide 1D nanostructures to use in different applications. The understanding of the interplay between OA crystal growth mechanism, phase transformation, and kinetics is the key to produce controlled 1D nanostructures, but the establishment of their relationship remains a tremendous challenge. The recent advances on in-situ, ex-situ techniques and theoretical simulations provide important insights into OA growth. However, the applicability of general kinetic models to interpret the OA mechanism was already questioned recently, as the growing interest of 1D nanostructures based on OA crystal growth. Therefore, the development of OA-based elongation kinetics of

1D nanostructures represents a key knowledge gap and a challenging task. Our recent time-dependent HR-TEM study provides new insights into crystal growth of $Cu(OH)_2$ nanowire formation from primary nanocrystals via OA mechanism in a sol-gel system. We introduced novel multistep sigmoidal kinetic models for the OA-based $Cu(OH)_2$ nanowire formation. This study significantly contributes to the advancement on the fabrication of ultrathin nanowires using OA attachment without using any surfactants by correlating OA-based elongation growth kinetics. We anticipate that the next decade will be an exciting time for both materials scientists and computational scientists to study crystal growth dynamics with state-of-the-art advanced instruments.

Acknowledgements

This work was performed at the Joint School of Nanoscience and Nanoengineering, a member of the Southeastern Nanotechnology Infrastructure Corridor (SENIC) and National Nanotechnology Coordinated Infrastructure (NNCI), supported by the NSF (grant ECCS-1542174). Financial support for this work was provided, in part, by the Joint School of Nanoscience and Nanoengineering, the University of North Carolina at Greensboro Office of Research and Sponsored Programs, National Science Foundation Science and Technologies for Phosphorus Sustainability (STEPS) (CBET-2019435). The authors acknowledge the DOD HBCU/MSI instrumentation award (contract #: W911NF1910522) to acquire HR-TEM (JEOL 2100PLUS) with STEM/EDS capability.

Conflict of interest

There is no conflict of interest to declare.

Author details

Gayani Pathiraja[1], Sherine Obare[1,2]* and Hemali Rathnayake[1]

1 Department of Nanoscience, Joint School of Nanoscience and Nanoengineering, University of North Carolina at Greensboro, Greensboro, NC, USA

2 Department of Nanoengineering, Joint School of Nanoscience and Nanoengineering, North Carolina A&T State University, Greensboro, NC, USA

*Address all correspondence to: sherine.obare@wmich.edu

References

[1] Cho K-S, Talapin DV, Gaschler W, Murray CB. Designing PbSe nanowires and nanorings through oriented attachment of nanoparticles. Journal of the American Chemical Society. 2005; **127**:7140-7147. DOI: 10.1021/ja050107s

[2] Zhang J, Huang F, Lin Z. Progress of nanocrystalline growth kinetics based on oriented attachment. Nanoscale. 2010;**2**: 18-34. DOI: 10.1039/B9NR00047J

[3] Murphy CJ, Jana NR. Controlling the aspect ratio of inorganic nanorods and nanowires. Advanced Materials. 2002; **14**:80-82

[4] Penn RL. Kinetics of oriented aggregation. The Journal of Physical Chemistry B. 2004;**108**:12707-12712. DOI: 10.1021/jp036490+

[5] Kuchibhatla SVNT, Karakoti AS, Bera D, Seal S. One dimensional nanostructured materials. Progress in Materials Science. 2007;**52**:699-913. DOI: 10.1016/j.pmatsci.2006.08.001

[6] Pathiraja G, Rathnayake H. Ultrathin metal hydroxide/oxide nanowires: Crystal growth, self-assembly, and fabrication for optoelectronic applications. In: Nanostructured Materials—Classification, Growth, Simulation, Characterization, and Devices. London, UK: IntechOpen; 2021

[7] Devan RS, Patil RA, Lin J-H, Ma Y-R. One-dimensional metal-oxide nanostructures: Recent developments in synthesis, characterization, and applications. Advanced Functional Materials. 2012;**22**:3326-3370. DOI: 10.1002/adfm.201201008

[8] Huang Y, Duan X, Wei Q, Lieber CM. Directed assembly of one-dimensional nanostructures into functional networks. Science. 2001;**291**:630-633. DOI: 10.1126/science.291.5504.630

[9] Murph SEH, Murphy CJ, Leach A, Gall K. A possible oriented attachment growth mechanism for silver nanowire formation. Crystal Growth & Design. 2015;**15**:1968-1974. DOI: 10.1021/acs.cgd.5b00123

[10] De Yoreo JJ, Chung S, Friddle RW. In situ atomic force microscopy as a tool for investigating interactions and assembly dynamics in biomolecular and biomineral systems. Advanced Functional Materials. 2013;**23**:2525-2538. DOI: 10.1002/adfm.201203424

[11] Shin S-H, Chung S, Sanii B, Comolli LR, Bertozzi CR, De Yoreo JJ. Direct observation of kinetic traps associated with structural transformations leading to multiple pathways of S-layer assembly. Proceedings of the National Academy of Sciences. 2012;**109**:12968-12973. DOI: 10.1073/pnas.1201504109

[12] Herrmann J, Li P-N, Jabbarpour F, Chan ACK, Rajkovic I, Matsui T, et al. A bacterial surface layer protein exploits multistep crystallization for rapid self-assembly. Proceedings of the National Academy Science USA. 2020;**117**: 388-394. DOI: 10.1073/pnas.1909798116

[13] Bhusari R, Thomann J-S, Guillot J, Leturcq R. Morphology control of copper hydroxide based nanostructures in liquid phase synthesis. Journal of Crystal Growth. 2021;**570**:126225. DOI: 10.1016/j.jcrysgro.2021.126225

[14] Zhu C, Liang S, Song E, Zhou Y, Wang W, Shan F, et al. In-situ liquid cell transmission electron microscopy investigation on oriented attachment of gold nanoparticles. Nature

Communications. 2018;**9**:421. DOI: 10.1038/s41467-018-02925-6

[15] Mujin M, Nakamura M, Matsumoto M. In-situ FE-SEM observation of the growth behaviors of fe particles at magmatic temperatures. Journal of Crystal Growth. 2021;**560**: 126043. DOI: 10.1016/j.jcrysgro.2021.126043

[16] Nakagawa Y, Kageyama H, Oaki Y, Imai H. Direction control of oriented self-assembly for 1D, 2D, and 3D microarrays of anisotropic rectangular nanoblocks. Journal of the American Chemical Society. 2014;**136**:3716-3719. DOI: 10.1021/ja410183q

[17] Lv W, He W, Wang X, Niu Y, Cao H, Dickerson JH, et al. Understanding the oriented-attachment growth of nanocrystals from an energy point of view: A review. Nanoscale. 2014;**6**: 2531-2547. DOI: 10.1039/C3NR04717B

[18] Niederberger M, Cölfen H. Oriented attachment and mesocrystals: Non-classical crystallization mechanisms based on nanoparticle assembly. Physical Chemistry Chemical Physics. 2006;**8**: 3271-3287. DOI: 10.1039/B604589H

[19] Penn RL. Imperfect oriented attachment: Dislocation generation in defect-free nanocrystals. Science. 1998; **281**:969-971. DOI: 10.1126/science.281.5379.969

[20] Ning J, Men K, Xiao G, Zou B, Wang L, Dai Q, et al. Synthesis of narrow band gap SnTe nanocrystals: Nanoparticles and single crystal nanowires via oriented attachment. CrystEngComm. 2010;**12**:4275. DOI: 10.1039/c004098n

[21] Penn RL, Banfield JF. Oriented attachment and growth, twinning, polytypism, and formation of metastable phases; Insights from nanocrystalline TiO2. American Mineralogist. 1998;**83**: 1077-1082. DOI: 10.2138/am-1998-9-1016

[22] Moldovan D, Yamakov V, Wolf D, Phillpot SR. Scaling behavior of grain-rotation-induced grain growth. Physical Review Letters. 2002;**89**:206101. DOI: 10.1103/PhysRevLett.89.206101

[23] Ribeiro C, Lee EJH, Giraldi TR, Longo E, Varela JA, Leite ER. Study of synthesis variables in the nanocrystal growth behavior of tin oxide processed by controlled hydrolysis. The Journal of Physical Chemistry. B. 2004;**108**: 15612-15617. DOI: 10.1021/jp0473669

[24] Cademartiri L, Ozin GA. Ultrathin nanowires: A materials chemistry perspective. Advanced Materials. 2009; **21**:1013-1020. DOI: 10.1002/adma.200801836

[25] Dresselhaus MS, Chen G, Tang MY, Yang RG, Lee H, Wang DZ, et al. New directions for low-dimensional thermoelectric materials. Advanced Materials. 2007;**19**:1043-1053. DOI: 10.1002/adma.200600527

[26] Maurer JHM, González-García L, Reiser B, Kanelidis I, Kraus T. Templated self-assembly of ultrathin gold nanowires by nanoimprinting for transparent flexible electronics. Nano Letters. 2016;**16**:2921-2925. DOI: 10.1021/acs.nanolett.5b04319

[27] Nikolaev KG, Ermolenko YE, Offenhäusser A, Ermakov SS, Mourzina YG. Multisensor systems by electrochemical nanowire assembly for the analysis of aqueous solutions. Frontiers in Chemistry. 2018;**6**:256. DOI: 10.3389/fchem.2018.00256

[28] Zhang J, Lin Z, Lan Y, Ren G, Chen D, Huang F, et al. A multistep

oriented attachment kinetics: Coarsening of ZnS nanoparticle in concentrated NaOH. Journal of the American Chemical Society. 2006;**128**:12981-12987. DOI: 10.1021/ja062572a

[29] Huang F, Banfield JF. Size-dependent phase transformation kinetics in nanocrystalline ZnS. Journal of the American Chemical Society. 2005;**127**: 4523-4529. DOI: 10.1021/ja048121c

[30] Chernov AA. Modern crystallography. In: Cardona M, Fulde P, Queisser H-J, editors. Springer Series in Solid-State Sciences. Berlin, Heidelberg: Springer Berlin Heidelberg; 1984. p. 36

[31] Ru-Shi L. Controlled Nanofabrication: Advances and Applications. Dubai, U.A.E.: Jenny Stanford Publishing; 2012

[32] Jun Y, Casula MF, Sim J-H, Kim SY, Cheon J, Alivisatos AP. Surfactant-assisted elimination of a high energy facet as a means of controlling the shapes of TiO2 nanocrystals. Journal of the American Chemical Society. 2003;**125**: 15981-15985. DOI: 10.1021/ja0369515

[33] Dinh C-T, Nguyen T-D, Kleitz F, Do T-O. Shape-controlled synthesis of highly crystalline titania nanocrystals. ACS Nano. 2009;**3**:3737-3743. DOI: 10.1021/nn900940p

[34] Halder A, Ravishankar N. Ultrafine single-crystalline gold nanowire arrays by oriented attachment. Advanced Materials. 2007;**19**:1854-1858. DOI: 10.1002/adma.200602325

[35] Yang Q, Lu Z, Liu J, Lei X, Chang Z, Luo L, et al. Metal oxide and hydroxide nanoarrays: Hydrothermal synthesis and applications as supercapacitors and nanocatalysts. Progress in Natural Science: Materials International. 2013;**23**: 351-366. DOI: 10.1016/j.pnsc.2013.06.015

[36] Deng Z, Chen D, Tang F, Meng X, Ren J, Zhang L. Orientated attachment assisted self-assembly of Sb2O3 nanorods and nanowires: End-to-end versus side-by-side. Journal of Physical Chemistry C. 2007;**111**:5325-5330. DOI: 10.1021/jp068545o

[37] Pacholski C, Kornowski A, Weller H. Self-assembly of ZnO: From nanodots to nanorods. Angewandte Chemie International Edition. 2002;**41**: 1188-1191. DOI: 10.1002/1521-3773(20020402)41:7%3C1188::AID-ANIE1188%3E3.0.CO;2-5

[38] Davis K, Yarbrough R, Froeschle M, White J, Rathnayake H. Band gap engineered zinc oxide nanostructures via a sol–gel synthesis of solvent driven shape-controlled crystal growth. RSC Advances. 2019;**9**:14638-14648. DOI: 10.1039/C9RA02091H

[39] Cao X, Wang N, Wang L. Ultrathin ZnO nanorods: Facile synthesis. Characterization and Optical Properties. Nanotechnology. 2010;**21**:065603. DOI: 10.1088/0957-4484/21/6/065603

[40] Chaurasiya N, Kumar U, Sikarwar S, Yadav BC, Yadawa PK. Synthesis of TiO2 nanorods using wet chemical method and their photovoltaic and humidity sensing applications. Sensors International. 2021;**2**:100095. DOI: 10.1016/j.sintl.2021.100095

[41] Jayachandran M, Rose A, Maiyalagan T, Poongodi N, Vijayakumar T. Effect of various aqueous electrolytes on the electrochemical performance of α-MnO2 nanorods as electrode materials for supercapacitor application. Electrochimica Acta. 2021;**366**:137412. DOI: 10.1016/j.electacta.2020.137412

[42] Sundar S, Venkatachalam G, Kwon S. Biosynthesis of copper oxide

(CuO) nanowires and their use for the electrochemical sensing of dopamine. Nanomaterials. 2018;**8**:823. DOI: 10.3390/nano8100823

[43] Zhang Y, Wang S, Li X, Chen L, Qian Y, Zhang Z. CuO shuttle-like nanocrystals synthesized by oriented attachment. Journal of Crystal Growth. 2006;**291**:196-201. DOI: 10.1016/j.jcrysgro.2006.02.044

[44] Pathiraja G, Yarbrough R, Rathnayake H. Fabrication of ultrathin CuO nanowires augmenting oriented attachment crystal growth directed self-assembly of Cu(OH)2 colloidal nanocrystals. Nanoscale Advances. 2020; **2**:2897-2906. DOI: 10.1039/D0NA00308E

[45] Liang H, Meng F, Lamb BK, Ding Q, Li L, Wang Z, et al. Solution growth of screw dislocation driven α-GaOOH nanorod arrays and their conversion to porous ZnGa 2 O 4 nanotubes. Chemistry of Materials. 2017;**29**: 7278-7287. DOI: 10.1021/acs.chemmater.7b01930

[46] Zhang X, He Y, Sushko ML, Liu J, Luo L, De Yoreo JJ, et al. Direction-specific van Der Waals attraction between Rutile TiO 2 nanocrystals. Science. 2017;**356**:434-437. DOI: 10.1126/science.aah6902

[47] Li D, Nielsen MH, Lee JR, Frandsen C, Banfield JF, De Yoreo JJ. Direction-specific interactions control crystal growth by oriented attachment. Science. 2012;**336**:1014-1018. DOI: 10.1126/science.1219643

[48] Leite ER, Giraldi TR, Pontes FM, Longo E, Beltran A, Andres J. Crystal growth in colloidal tin oxide nanocrystals induced by coalescence at room temperature. Applied Physics Letters. 2003;**83**:1566-1568. DOI: 10.1063/1.1605241

[49] Hapiuk D, Masenelli B, Masenelli-Varlot K, Tainoff D, Boisron O, Albin C, et al. Oriented attachment of ZnO nanocrystals. Journal of Physical Chemistry C. 2013;**117**:10220-10227. DOI: 10.1021/jp402738y

[50] Jin B, Sushko ML, Liu Z, Jin C, Tang R. In Situ liquid cell TEM reveals bridge-induced contact and fusion of Au nanocrystals in aqueous solution. Nano Letters. 2018;**18**:6551-6556. DOI: 10.1021/acs.nanolett.8b03139

[51] Niu W, Zhang L, Xu G. Seed-mediated growth method for high-quality noble metal nanocrystals. Science China. Chemistry. 2012;**55**:2311-2317. DOI: 10.1007/s11426-012-4681-z

[52] Xia Y, Gilroy KD, Peng H-C, Xia X. Seed-mediated growth of colloidal metal nanocrystals. Angewandte Chemie, International Edition. 2017;**56**:60-95. DOI: 10.1002/anie.201604731

[53] Busbee BD, Obare SO, Murphy CJ. An improved synthesis of high-aspect-ratio gold nanorods. Advanced Materials. 2003;**15**:414-416. DOI: 10.1002/adma.200390095

[54] Zhang H, Penn RL, Lin Z, Cölfen H. Nanocrystal growth via oriented attachment. CrystEngComm. 2014;**16**: 1407. DOI: 10.1039/c4ce90001d

[55] Kirchner HOK. Coarsening of grain-boundary precipitates. MT. 1971;**2**: 2861-2864. DOI: 10.1007/BF02813264

[56] Xue X, Penn RL, Leite ER, Huang F, Lin Z. Crystal growth by oriented attachment: Kinetic models and control factors. CrystEngComm. 2014;**16**:1419. DOI: 10.1039/c3ce42129e

[57] Sangwal K. Nucleation and Crystal Growth: Metastability of Solutions and Melts. Hoboken, NJ, USA: John Wiley &

Sons, Inc.; 2018. DOI: 10.1002/9781119461616

[58] Zhang J, Wang Y, Zheng J, Huang F, Chen D, Lan Y, et al. Oriented attachment kinetics for ligand capped nanocrystals: Coarsening of Thiol-PbS nanoparticles. The Journal of Physical Chemistry. B. 2007;**111**:1449-1454. DOI: 10.1021/jp067040v

[59] Zhuang Z, Zhang J, Huang F, Wang Y, Lin Z. Pure multistep oriented attachment growth kinetics of surfactant-Free SnO2 nanocrystals. Physical Chemistry Chemical Physics. 2009;**11**:8516. DOI: 10.1039/b907967j

[60] Madras G, McCoy BJ. Distribution kinetics theory of Ostwald ripening. The Journal of Chemical Physics. 2001;**115**: 6699-6706. DOI: 10.1063/1.1403687

[61] Huang F, Zhang H, Banfield JF. Two-stage crystal-growth kinetics observed during hydrothermal coarsening of nanocrystalline ZnS. Nano Letters. 2003;**3**(3):373-378. DOI: 10.1021/nl025836+

[62] Joesten R, Fisher G. Kinetics of diffusion-controlled mineral growth in the Christmas Mountains (Texas) Contact Aureole. GSA Bulletin. 1988; **100**:714-732. DOI: 10.1130/0016-7606(1988)100<0714:KODCMG>2.3.CO;2

[63] Xiao J, Qi L. Surfactant-assisted, shape-controlled synthesis of gold nanocrystals. Nanoscale. 2011;**3**: 1383-1396. DOI: 10.1039/C0NR00814A

[64] Kumar CS. Transmission Electron Microscopy Characterization of Nanomaterials. Berlin/Heidelberg, Germany: Springer Science & Business Media; 2013

[65] Smith DJ. Chapter 1. Characterization of nanomaterials using transmission electron microscopy. In: Kirkland AI, Haigh SJ, editors. Nanoscience & Nanotechnology. Cambridge: Royal Society of Chemistry; 2015. pp. 1-29. DOI: 10.1039/9781782621867-00001

[66] Nellist PD. The principles of STEM imaging. In: Pennycook SJ, Nellist PD, editors. Scanning Transmission Electron Microscopy. New York, NY: Springer New York; 2011. pp. 91-115

[67] Wang ZL. Transmission electron microscopy of shape-controlled nanocrystals and their assemblies. The Journal of Physical Chemistry. B. 2000; **104**:1153-1175. DOI: 10.1021/jp993593c

[68] Thangala J, Chen Z, Chin A, Ning C-Z, Sunkara MK. Phase transformation studies of metal oxide nanowires. Crystal Growth & Design. 2009;**9**:3177-3182. DOI: 10.1021/cg801198p

[69] Shi F, Li F, Ma Y, Zheng F, Feng R, Song C, et al. In situ transmission electron microscopy study of nanocrystal formation for electrocatalysis. ChemNanoMat. 2019;**5**: 1439-1455. DOI: 10.1002/cnma.201900497

[70] de Jonge N, Houben L, Dunin-Borkowski RE, Ross FM. Resolution and aberration correction in liquid cell transmission electron microscopy. Nature Reviews Materials. 2019;**4**:61-78. DOI: 10.1038/s41578-018-0071-2

[71] Nielsen MH, Lee JRI, Hu Q, Yong-Jin Han T, De Yoreo JJ. Structural evolution, formation pathways and energetic controls during template-directed nucleation of CaCO3. Faraday Discussion. 2012;**159**:105

[72] Nielsen MH, Aloni S, De Yoreo JJ. In Situ TEM imaging of CaCO3 nucleation reveals coexistence of direct and indirect

pathways. Science. 2014;**345**:1158-1162. DOI: 10.1126/science.1254051

[73] Liu Y, Lin X-M, Sun Y, Rajh T. In situ visualization of self-assembly of charged gold nanoparticles. Journal of the American Chemical Society. 2013; **135**:3764-3767. DOI: 10.1021/ja312620e

[74] Woehl T. Refocusing in situ electron microscopy: Moving beyond visualization of nanoparticle self-assembly to gain practical insights into advanced material fabrication. ACS Nano. 2019;**13**:12272-12279. DOI: 10.1021/acsnano.9b08281

[75] Jin B, Liu Z, Tang R. Recent experimental explorations of non-classical nucleation. CrystEngComm. 2020;**22**:4057-4073. DOI: 10.1039/D0CE00480D

[76] Liu C, Ou Z, Zhou S, Chen Q. Nonclassical crystallization observed by liquid-phase transmission electron microscopy. In: Zhang X, editor. ACS Symposium Series. Washington, DC: American Chemical Society; 2020

[77] Wei W, Zhang H, Wang W, Dong M, Nie M, Sun L, et al. Observing the growth of Pb 3 O 4 nanocrystals by in situ liquid cell transmission electron microscopy. ACS Applied Materials & Interfaces. 2019;**11**:24478-24484. DOI: 10.1021/acsami.9b08524

[78] Song M, Zhou G, Lu N, Lee J, Nakouzi E, Wang H, et al. Oriented attachment induces fivefold twins by forming and decomposing high-energy grain boundaries. Science. 2020;**367**: 40-45. DOI: 10.1126/science.aax6511

[79] Liu L, Nakouzi E, Sushko ML, Schenter GK, Mundy CJ, Chun J, et al. Connecting energetics to dynamics in particle growth by oriented attachment using real-time observations. Nature Communications. 2020;**11**:1045. DOI: 10.1038/s41467-020-14719-w

[80] Guinier A, Fournet G, Yudowitch KL. Small-Angle Scattering of X-Rays. Hoboken, New Jersey: John Wiley & Sons; 1955. DOI: 10.1002/pol.1956.120199326

[81] Abécassis B, Testard F, Spalla O, Barboux P. Probing in situ the nucleation and growth of gold nanoparticles by small-angle X-ray scattering. Nano Letters. 2007;**7**:1723-1727. DOI: 10.1021/nl0707149

[82] Polte J, Ahner TT, Delissen F, Sokolov S, Emmerling F, Thünemann AF, et al. Mechanism of gold nanoparticle formation in the classical citrate synthesis method derived from coupled In Situ XANES and SAXS evaluation. Journal of the American Chemical Society. 2010;**132**: 1296-1301. DOI: 10.1021/ja906506j

[83] Polte J, Erler R, Thünemann AF, Sokolov S, Ahner TT, Rademann K, et al. Nucleation and growth of gold nanoparticles studied via in Situ small angle X-ray scattering at millisecond time resolution. ACS Nano. 2010;**4**: 1076-1082. DOI: 10.1021/nn901499c

[84] Tsao C-S, Chuang C-M, Chen C-Y, Huang Y-C, Cha H-C, Hsu F-H, et al. Reaction kinetics and formation mechanism of TiO 2 nanorods in solution: An insight into oriented attachment. Journal of Physical Chemistry C. 2014;**118**:26332-26340. DOI: 10.1021/jp506780h

[85] Chen R, Nguyen QN, Xia Y. Oriented attachment: A unique mechanism for the colloidal synthesis of metal nanostructures. ChemNanoMat. 2022;**8**:1-22. DOI: 10.1002/cnma.202100474

[86] Zhang H, Banfield JF. Interatomic Coulombic interactions as the driving force for oriented attachment. CrystEngComm. 2014;**16**:1568-1578. DOI: 10.1039/C3CE41929K

[87] Song H, Lee K-H, Jeong H, Um SH, Han G-S, Jung HS, et al. A simple self-assembly route to single crystalline SnO2 nanorod growth by oriented attachment for dye sensitized solar cells. Nanoscale. 2013;**5**:1188. DOI: 10.1039/c2nr33114d

[88] Ghezelbash A, Koo B, Korgel BA. Self-assembled stripe patterns of CdS nanorods. Nano Letters. 2006;**6**: 1832-1836. DOI: 10.1021/nl061035l

[89] Zhu G, Zhang S, Xu Z, Ma J, Shen X. Ultrathin ZnS single crystal nanowires: Controlled synthesis and room-temperature ferromagnetism properties. Journal of the American Chemical Society. 2011;**133**:15605-15612. DOI: 10.1021/ja2049258

[90] Tang Z. Spontaneous organization of single CdTe nanoparticles into luminescent nanowires. Science. 2002; **297**:237-240. DOI: 10.1126/ science.1072086

[91] Zhang X, Shen Z, Liu J, Kerisit SN, Bowden ME, Sushko ML, et al. Direction-specific interaction forces underlying zinc oxide crystal growth by oriented attachment. Nature Communications. 2017;**8**:835. DOI: 10.1038/s41467-017-00844-6

[92] Spagnoli D, Banfield JF, Parker SC. Free energy change of aggregation of nanoparticles. Journal of Physical Chemistry C. 2008;**112**:14731-14736. DOI: 10.1021/jp804966c

[93] Stroppa DG, Montoro LA, Beltrán A, Conti TG, da Silva RO, Andrés J, et al. Unveiling the chemical and morphological features of Sb–SnO 2 nanocrystals by the combined use of high-resolution transmission electron microscopy and Ab Initio surface energy calculations. Journal of the American Chemical Society. 2009;**131**: 14544-14548. DOI: 10.1021/ja905896u

[94] Faccin GM, Pereira ZS, da Silva EZ. How crystallization affects the oriented attachment of silver nanocrystals. Journal of Physical Chemistry C. 2021; **125**:6812-6820. DOI: 10.1021/acs. jpcc.0c10321

[95] Sayle TXT, Inkson BJ, Karakoti A, Kumar A, Molinari M, Möbus G, et al. Mechanical properties of ceria nanorods and nanochains; the effect of dislocations. Grain-Boundaries and Oriented Attachment. Nanoscale. 2011;**3**: 1823. DOI: 10.1039/c0nr00980f

[96] Zhang Q, Liu S-J, Yu S-H. Recent advances in oriented attachment growth and synthesis of functional materials: Concept, evidence, mechanism, and future. Journal of Materials Chemistry. 2009;**19**:191-207. DOI: 10.1039/ B807760F

[97] Tsai C-C, Teng H. Regulation of the physical characteristics of Titania Nanotube aggregates synthesized from hydrothermal treatment. Chemistry of Materials. 2004;**16**:4352-4358. DOI: 10.1021/cm049643u

[98] Nian J-N, Teng H. Hydrothermal synthesis of single-crystalline anatase TiO 2 nanorods with nanotubes as the precursor. The Journal of Physical Chemistry. B. 2006;**110**:4193-4198. DOI: 10.1021/jp0567321

[99] Finnegan MP, Zhang H, Banfield JF. Anatase coarsening kinetics under hydrothermal conditions as a function of Ph and temperature. Chemistry of Materials. 2008;**20**:3443-3449. DOI: 10.1021/cm071057o

[100] Gunning RD, O'Sullivan C, Ryan KM. A multi-rate kinetic model for spontaneous oriented attachment of CdS nanorods. Physical Chemistry Chemical Physics. 2010;**12**:12430-12435. DOI: 10.1039/C0CP00196A

[101] He W, Lin J, Lin X, Lu N, Zhou M, Zhang KHL. The evaluation of Coulombic interaction in the oriented-attachment growth of colloidal nanorods. Analyst. 2012;**137**:4917. DOI: 10.1039/c2an35950b

[102] He W, Lin J, Wang B, Tuo S, Pantelides ST, Dickerson JH. An analytical expression for the van Der Waals interaction in oriented-attachment growth: A spherical nanoparticle and a growing cylindrical nanorod. Physical Chemistry Chemical Physics. 2012;**14**:4548. DOI: 10.1039/C2CP23919A

[103] Pathiraja G, Herr DJC, Rathnayake H. Nanoscopic insight into Sol–Gel chemical kinetics of oriented attachment crystal growth in anisotropic copper hydroxide nanowires. Crystal Growth & Design. 2022;**22**:2889-2902. DOI: 10.1021/acs.cgd.1c01247

[104] Wang Y, Peng X, Abelson A, Zhang B-K, Qian C, Ercius P, et al. In Situ TEM observation of neck formation during oriented attachment of PbSe nanocrystals. Nano Research. 2019;**12**: 2549-2553. DOI: 10.1007/s12274-019-2483-8

Chapter 5

Recent Advances in Infrared Nonlinear Optical Crystal

Senthil Kumar Chandran, Chinnakannu Elavarasi, Srinivasan Manikam and John James Gnanapragasam

Abstract

The search and growth of nonlinear optical (NLO) crystals in the infrared (IR) area are significant and of high importance in the fields of NLO, signal communication, solid-state chemistry, and laser frequency conversion. Infrared NLO crystals have a wide IR transparent range, high laser damage threshold (LDT) value, and large NLO coefficients. This chapter presents the recent advances in IR-NLO crystals and especially emphasizes their crystal growth method, crystal structures, band gap value, LDT, and NLO properties. Based on its structural variety, it is categorized into chalcogenides, chalcohalides, oxides, halides, and oxyhalides. This chapter describes several kinds of IR-NLO crystals and their structural, band gap value, thermal, optical, LDT, and NLO properties and also describes the significance of these crystals in laser frequency conversion, optical parameter oscillator, and other optical applications.

Keywords: crystal growth, laser-induced damage thresholds, infrared crystal, transmittance, optical parameter oscillator

1. Introduction

Nonlinear optical (NLO) crystals are important for frequency conversion and are widely used in different laser-oriented applications. High-efficiency NLO crystals need in high-efficiency laser methods, so it is essential to growing novel NLO crystals with good properties. In the past five decades, many valuable NLO crystals in the near-infrared (NIR), visible, and ultraviolet areas have been commercialized, such as $LiNbO_3$, LiB_3O_5 (LBO), β-BaB_2O_4 (β-BBO), $KTiOPO_4$ (KTP), and KH_2PO_4 (KDP). These crystals are commonly used in basic science and technology, such as laser generation, artificial nuclear fusion, and so on. However, due to the increasing practical or market necessities, only a few crystals can be successfully used in deep-UV(DUV) and mid/far-IR areas. Nevertheless, NLO crystals that can powerfully produce high-power mid-IR lasers in the spectral area of 2–25 μm are very rare. Up to now, several useful NLO crystals have been originated and used in DUV, UV-Vis, and near IR, but they cannot be implemented in mid-infrared spectra because they have two atmospheric transparent regions, 3–5 and 8–14 mm, owing to strong absorption [1–5].

Second-order nonlinear optical (NLO) crystals are significant for producing coherent energy in the IR region (3–20 μm). IR lasers have several vital applications in various devices, such as optical parametric oscillators (OPO), remote sensing, optical sensing, instrumental spectroscopy, industry, military, analytical devices optical imaging, laser guidance, telecommunication, medical diagnostics, and long-distance communications. Such instruments are used to identify various elements and precise vibrational spectra [6–9]. Even after several years of deep research, only three unresolved NLO crystals have been commercially accessible in the mid and far-IR areas, namely, $AgGaSe_2$, $AgGaS_2$, and $ZnGeP_2$. However, some inadequacies still exist in these IR-NLO crystals, such as the inherent efficiency loss arising from dual photon absorption, low MIR cut-off edge, non-phase matchable behavior, and low laser damage threshold. Already commercial ones cannot meet the commercial conditions because of their inherent disadvantages. Hence, it is essential to find out new efficient MIR NLO crystals with more stable efficiency [3, 7, 8, 10].

The large size high-quality mid-IR-NLO crystals for laser device applications are grown by Bridgman-Stockbarger (BS) method and molecular beam epitaxy (MBE) growth method [5]. The IR-NLO crystal can be separated into five categories: chalcogenides, chalcohalides, oxides, oxyhalides, halides, and chalcohalides [4]. This chapter will emphasize second-order NLO inorganic crystals in the MIR region. We did not focus on the commercially accessible $LiNbO_3$, LiB_3O_5 (LBO), β-BaB_2O_4 (β-BBO), KH_2PO_4, and $KTiOPO_4$ NLO crystals, these crystals have absorption bands in this region. Instead, this chapter focus on the chalcogenides, chalcohalides, oxides, halides, and oxyhalides. These crystals are promising materials for MIR applications due to they have wide transmittance in the MIR region [11].

2. Main conditions at NLO crystal select for changing coherent energy in the IR region

The selection of high-quality crystals is mainly tough when designing new, appropriate NLO crystals for the IR region. It should be highlighted that the balance between SHG coefficients and energy gap is an important feature to attain noble optical functioning in a mid-IR-NLO crystal [3]. The mid-IR-NLO crystals are significant to develop high-power tunable laser output extending the two atmospheric bands (3–5 μm and 8–14 μm) [4]. The accessibility of bulk-size single crystals is vital for the production of NLO devices. It is a great task to grow novel mid-IR-NLO crystals with desirable properties for useful applications. The good mid-IR-NLO crystals should satisfy the following basic criteria [2, 3, 6, 10, 12, 13]:

1. High second harmonic generation (SHG) responses

2. High laser damage threshold

3. Wide IR optical transparency range

4. Moderate birefringence

5. Good thermal stability

6. Good mechanical properties

DOI: http://dx.doi.org/10.5772/intechopen.108173

7. Good facile growth of big size crystals

8. Splendid chemical stability

9. Crystal with non-centrosymmetric space group

10. Congruent-melting performance to enable single-crystal growth

11. Low absorption loss at suitable laser wavelengths

3. Chalcogenides

Chalcogenides are new kind of material denoting the chalcogen elements (sulfide, selenide, and telluride) of the group VIA. The chalcogenides are designed by covalent bonding and originate in a variety of structures, mostly formatted in octahedral or trigonal geometry. Chalcogenides are useful in many fields, such as photocatalyst, thermoelectric, MIR-NLO, photovoltaic, sensor, fuel cell, and battery [6, 14, 15]. Chalcogenides are suitable crystals for MIR-NLO as they exhibit wide transparency in IR regions and can obtain large SHG responses in this region [16]. Normally, chalcogenides are capable materials for MIR-NLO devices owing to their many benefits, such as large optical nonlinearity, broad transparency range, and large birefringence. II-IV-V_2 and I-III-VI_2 chalcopyrites are now the leading functional MIR-NLO crystals in the market and laboratory. In the past two decades, more consideration has been given to discovering chalcogenides as MIR-NLO crystals for their structural diversity. Quaternary chalcogenide crystals own a high bandgap (Eg) and high LDT. But the small nonlinearity coefficients slowed down their use in high-power laser generation. Though such kinds of crystals have many MIR-NLO benefits, many of them also have some inadequacies. To meet the requirement of laser device manufacture, some advanced growth methods are adopted to produce high purity and high-quality big size crystals of the chalcogenide. But, quaternary chalcogenide crystals, such as $Li_2Ga_2GeS_6$, $LiGaGe2Se_6$, $AgGaGeS_4$, and Ba_2GaGeS_6, are grown in bulk-sized crystals, which is desirable for optical devices [5, 6, 10, 16]. There are different kinds of chalcogenides, generally, alkali metal chalcogenides and transition metal chalcogenides (TMCs), which again can be categorized into binary, ternary, and quaternary chalcogenides. The chalcogenides have weaker interatomic bonds than the oxides, resulting in good optical transparencies in the IR regions. Meanwhile, the chalcogenides exhibit adjustable structure and optical properties [6, 14, 15, 17].

3.1 Cataloguing of chalcogenides based on number of components

Chalcogenides are classified based on their number of components, such as binary, ternary and quaternary structures, number of metals, number of chalcogen ions, and so on though, both ternary and quaternary elements of chalcogenides are systematically analyzed compared to binary chalcogenides [18, 19].

3.1.1 Binary chalcogenides

Binary chalcogenides containes two kinds of ions (metal ions and chalcogen anion). The CdS, CdSe, Ga_2S_3, GaS, In_2S_3, GaSe MnS, SnS, SnS_2, ZnS, and ZnSe are an example of binary chalcogenides. For instance, CdS is one of the most considered chalcogenides. It has an energy gap value of 2.3 eV and is comparatively active under visible light. Its special size and structure-based optical and electronic properties are desirable for various kinds of applications. Owing to its numerous possible applications CDS chalcogenides are assumed to be the most significant materials [18, 19].

3.1.2 Ternary chalcogenides

When selecting a crystal for laser energy conversion in the IR, it is essential to have an ideal mixture of various considerations like birefringence value not less than 0.03 and LDT value of around 100 Mcm^{-2}, energy gap value of more than 3.3 eV and the NLO coefficient should be more than 4 pm V^{-1} [20]. The Li- and Ba-having chalcogenides meet these desires. The $ZnGeP_2$, $AgGaSe_2$, $CdSiP_2$, and $AgGaTe_2$ crystals have high NLO susceptibility, but their forbidden energy band is too low. The most commonly used nonlinear crystals for the MIR are $AgGaS_2$ $AgGaSe_2$, and $ZnGeP_2$. However, they all own serious disadvantages [6, 21, 22]. In recent times, consideration was given to chalcogenide crystals, such as alkali and alkali-earth metals (Li and Ba) (**Table 1**). These crystals permit one to resolve some difficulties in the MIR region. The birefringence value in Li-comprising crystals is significantly larger. $LiBC_2$ (B = Ga; C = S,Se) crystals can be applied for SHG applications wavelengths between 1.4 and 12 μm [23, 24]. Telluride crystals also have MIR properties, especially $LiGaTe_2$ has phase-matching in the entire transparency region. The SHG conversion efficiency of $LiGaTe_2$ is 10.6 μm, which is higher than that of AGSe [25]. To enhance the energy gap, Ag cation is to be substituted with alkali/alkaline earth metal (Li, Ba). Adding

Crystal	Point group	Transparency range (μm)	Band gap (ev)	Nonlinear coeff. (pm V¯1)	Laser damage threshold (MW cm¯2)
$AgGaS_2$	42 m	0.47–13	2.7	d32 = 8, 1.06 d36 = 19, 1.06	34 at 10 ns, 1064 nm
$ZnGeP_2$	42 m	0.74–12	2	d36=75, 9.6	100 at 10 ns, 2000 nm
$AgGaSe_2$	42 m	0.76–18	1.8	d32 = 19.6,3.4 d36 = 39, 1.06	13 at 30 ns, 2000 nm
$LiGaTe_2$	42 m	0.52–20	2.31	d36 = 43, 4.5	—
$LiGaS_2$	mm2	0.32–12	4.15	d31 = 5.8, 2.3 d24 = 5.1, 2.3	240 at 14 ns, 1064 nm
$LiGaSe_2$	mm2	0.37–14	3.57	d31 = 9.9,2.3 d24 = 7.7, 2.3	80 at 5.6 ns, 1064 nm
$LiInS_2$	mm2	0.34–13.2	3.57	d31 = 7.25 d24 = 5.66, 2.3	40* 14 ns, 1064 nm
$LiInSe_2$	mm2	0.46–14	2.86	d31 = 11.78 d24 = 8.17, 2.3	40* 10 ns, 1064 nm
$BaGa_4S_7$	mm2	0.35–13.7	3.54	d32 = 5.7, 2.3	250 at 14 ns, 1064 nm
$BaGa_4Se_7$	m	0.47–18	2.64	d11 = 18.2, 2.3 d13 = −0.6, 2.3	—

Table 1.
Point group, transparency range, band gap, and LDT value of ternary chalcogenides.

Figure 1.
Single crystals of $BaGa_4S_7$.

these metal, we can get $LiBC_2$ (B = In, Ga; C = S, Se, Te) and BaGa4C7 (C = S, Se) group crystals. These crystals own a high bandgap value [26–28]. A little mass of Li is the reason for high thermal conductivity and high vibrational frequencies. The thermal conductivity of Li mixtures is around five times higher than that of AGS (Se) [29] and four to eight times more than that of $BaGa_4S_7$ (Se) (**Figure 1**) [6, 28]. The laser damage threshold for LGS is 3.5 Jcm^{-2}, which is five times larger than LISe [30]. A similar result has been obtained when the Ag ion is substituted with Ba. $BaGa_4S_7$ crystal has high LDT and NLO susceptibility. Though, Ba cation slightly drops the band gap value. The point group is mm_2 and m for $BaGa_4S_7$ (**Figure 1**) and $BaGa_4Se_7$, respectively. Wide transparency regions of 0.35–12 and 0.47–15.0 μm, and energy gaps of 3.54 and 2.64 eV were found for both $BaGa_4S_7$ and $BaGa_4Se_7$ crystals, respectively. Both crystals have strong absorption peak at 15 μm. The $BaGa_4Se_7$ showed high nonlinear susceptibility of d_{11} = 18.2 pm V^{-1} [6, 31, 32].

3.1.3 Quaternary chalcogenides

Quaternary chalcogenides have four kinds of ions together with a chalcogen anion. These kinds of materials have different applications, such as MIR-NLO, solar cell absorbers, photocatalysts, and so on. Quaternary materials adopt different kinds of elements, which permits comparatively complex structural, electronic, and optical properties [6, 33]. Using the quaternary crystals, the IR-NLO parameters can be enhanced with a high content of NLO-active parts [6]. A compact organization of the microscopic NLO-active parts increases high macroscopic NLO outcomes [6]. The birefringence value of $AgGaSe_2$ is 0.05 and this value for $AgGaGe_3Se_8$ is 0.11. The enhanced LDT value of $AgGaGeS_4$ shows that it is a potential alternative crystal to the generally used $AgGaS_2$ for IR-NLO applications. $Li_2In_2SiSe_6$, $Li_2In_2GeSe_6$, $Li_2Ga_2GeS_6$, and $LiGaGe_2Se_6$ are the Li-having quaternary chalcogenides crystals (**Table 2**) [6, 34–37]. All these crystals have non-centrosymmetric crystal structures. $Li_2Ga_2GeS_6$ (**Figure 2**), $LiGaGe_2Se_6$ (**Figure 3**) crystals have orthorhombic crystal systems with space group Fdd_2 and $Li_2In_2GeSe_6$, Li_2In_2SiSe6 crystals own monoclinic crystal systems with space group Cc. Ba-having quaternary $BaGa_2GeS_6$, $BaGa_2GeSe_6$ structures, which are promising crystals for NLO applications [37]. The NLO susceptibilities of $Li_2Ga_2GeS_6$ are 16 pm/V, which is significantly higher than $LiGaS_2$ (5.8 pm/V). A similar result was noted for $LiGaGe_2Se_6$. $BaGa_2GeS_6$ and $BaGa_2GeSe_6$ crystals also have improved NLO parameters. The SHG experiments showed that both materials have phase-matched behavior. The calculated SHG coefficient is ~2.1 and

Crystal	Point group	Transparency range (μm)	Band gap (ev)	Nonlinear coeff., (pm V^{-1})	Optical damage threshold (MW cm^{-2})
$AgGaGeS_4$	mm2	0.42–12	2.8	d31 = 15, 1.06	50 at 15 ns, 1064 nm
$AgGaGe_3Se_8$	mm2	0.6–18	2.4	d31 = 33.4	–
$Li_2Ga_2GeS_6$	mm2	0.35–14	2.51	deff = 16, 1.06	>50 at 15 ns, 1064 nm
$LiGaGe_2Se_6$	mm2	0.47–18	2.64	d15 = 18.6, 2.09	50 at 10 ns, 1064 nm
$Li_2In_2GeS_6$	m	0.36	3.45	≈d36 = 12.6, 10.6	–
$Li_2In_2GeSe_6$	m	0.54	2.30	≈d36 AGSe	–
$Li_2In_2SiS_6$	m	0.34	3.61	≈d36 AGS	–
$BaGa_2GeS_6$	3	0.38–14	3.26	deff = 26.3, 2.09	–
$BaGa_2GeSe_6$	3	0.44–18	2.81	deff = 43.7, 2.09	–

Table 2.
Point group, transparency range, band gap, and LDT value of quaternary chalcogenides.

Figure 2.
Crystal of $LiGaGe_2Se_6$.

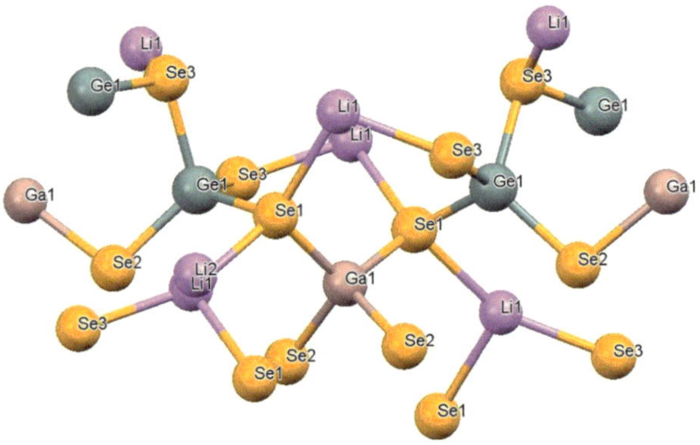

Figure 3.
Crystal structure of $LiGaGe_2Se_6$.

~3.5 times higher than that of $AgGaS_2$. The nonlinear susceptibilities are 26.3 pm/V and 43.7 pm/V for $BaGa_2GeS_6$ and $BaGa_2GeSe_6$, respectively [37]. The transparency region in $Li_2Ga_2GeS_6$ and $LiGaGe_2Se_6$ is 0.35–12 and 0.37–14 μm, respectively. For $BaGa_2GeS_6$ and $BaGa_2GeSe_6$, the transparency regions are 0.380–13.7 μm and 0.44–18 μm, respectively. Band gaps of $BaGa_2GeS_6$ and $BaGa_2GeSe_6$ are 3.26 and 2.81 eV, respectively. $Li_2In_2GeSe_6$ and $Li_2In_2SiSe_6$ crystals have the energy gap values of 2.30 and 3.61 eV, respectively [6].

4. Chalcohalides

Chalcohalides could be considered potential materials for mid-IR applications. The chalcohalides contain a combination of sulfur and halogen. Chalcohalides have non-centrosymmetric coordinated surroundings that can stimulate mid-IR-NLO efficiency. Based on the chemical compounds present in the chalcohalides, it can be divided into [4]:

1. Alkali/alkaline-earth metal chalcohalides
2. Adduct-type chalcohalides
3. Lewis acid adducts chalcohalides
4. Main group element clusters chalcohalides
5. Other group main metal chalcohalides
6. Transition metal chalcohalides

4.1 Alkali/alkaline-earth metal chalcohalides

The $Ba_4Ge_3S_9Cl_2$ chalcohalide has excellent mid-IR-NLO properties [38]. It has a space group $P6_3$. The bandgap energy of $Ba_4Ge_3S_9Cl_2$ was 2.91 eV. The SHG response of $Ba_4Ge_3S_9Cl_2$ was 2.4 times higher than that of AGS [38]. The bandgap of $Ba_4Ge_3S_9Cl_2$ is 2.67 eV, which was calculated by the DFT method. NLO coefficients are found to be $d_{15} = d_{24} = 7.61$ and $d_{33} = 13.81$ pm/V. The bandgap of $NaBa_4Ge_3S_{10}Cl$ was 3.49 eV and SHG efficiency was 0.3 times that of $AgGaS_2$ (AGS) [39]. The theoretically calculated bandgap of this crystal is 2.94 eV. Then, the NLO coefficients are calculated to be $d_{15} = d_{24} = 3.89$ and $d_{33} = 9.32$ pm/V. The mid-IR-NLO crystals $[A_3X][Ga_3PS_8]$ (A = K, Rb; X = Cl, Br) were synthesized and reported by B.W. Liu et al. in the year 2016 [40]. Two Cl-crystals $[K_3Cl][Ga_3PS_8]$ and $[Rb_3Cl][Ga_3PS_8]$ are having isostructural with space group $Pmn2_1$, while the other two Br-crystals $[Rb_3Br][Ga_3PS_8]$ and $[K_3Br][Ga_3PS_8]$ belong to Pm space group. These four compounds showed outstanding mid-IR-NLO behavior and the energy gaps of $[K_3Cl][Ga_3PS_8]$, $[Rb_3Cl][Ga_3PS_8]$, $[K_3Br][Ga_3PS_8]$, and $[Rb_3Br][Ga_3\text{-}PS_8]$ are 3.60, 3.65, 3.85, and 3.50 eV, respectively. All four compounds showed large SHG responses of 4.0, 5.0, 7.0, and 9.0 times that of $AgGaS_2$ (AGS) at 1064 nm [39]. Moreover, these four compounds have higher laser threshold damage (LDT) of 37, 35, 31, and 29 times than that of AGS (**Table 3**).

Crystal	Space group	Band gap (ev) (DFT)	SHG $AgGaS_2$ (AGS) times	Nonlinear coefficients (pm/V)	Laser threshold damage (LDT) $AgGaS_2$ (AGS)
$Ba_4Ge_3S_9Cl_2$	$P6_3$	2.67	2.4	$d_{15} = d_{24} = 7.61$ and $d_{33} = 13.81$	–
$NaBa_4Ge_3S_{10}Cl$	—	3.49	0.3	are $d_{15} = d_{24} = 3.89$ and $d_{33} = 9.32$	–
$[K_3Cl][Ga_3PS_8]$	Pmn21	3.60	4.0	–	37
$[Rb_3Cl][Ga_3PS_8]$	Pmn21	3.65	5.0	–	35
$[K_3Br][Ga_3PS_8]$	Pm	3.85,	7.0	–	35
$[Rb_3Br][Ga_3PS_8]$	Pm	3.50	9.0	–	29

Table 3.
Space group, band gap, and LDT value of alkali/alkaline-earth metal chalcohalides.

Crystal	Space group	Band gap (ev)	SHG $AgGaS_2$(AGS) times	IR transparent ranges (μm)	NLO coefficients (pm/V)
(SbI_3) $(S8)_3$	R3m	2.52	1.0	2.5–25.0	$d_{15} = 9.21$, $d_{22} = 9.22$, $d_{33} = 6.91$
(AsI_3). $(S_8)_3$	R3m	2.31 eV	0.8	0.4–25.0	$d_{15} = 3.40$, $d_{22} = 6.21$, $d_{33} = 0.73$
(SnI_4). $(S_8)_2$	Fdd_2	2.17 eV	0.5	2.5–25.0	–

Table 4.
Space group, band gap, SHG, and transparency range value of adduct-type chalcohalides.

4.2 Adduct-type chalcohalides

The (SbI_3) $(S8)_3$ and (SnI_4)-$(S8)_2$ are the adduct-type of chalcohalides. The (TI_3). $(S_8)_3$ (T = As, Sb) [4, 41–44] adduct-type chalcohalides have isostructural with R3m space group (**Table 4**). The (SbI_3). $(S_8)_3$ and (AsI_3). $(S_8)_3$ chalcohalides have reasonable energy gap value of 2.52 and 2.31 eV and their theoretical energy gaps are 2.69 and 2.21 eV, respectively. SHG responses of these two chalcohalides have 1.0 and 0.8 times that of $AgGaS_2$. They also have wide IR transparent wavelengths around 2.5–25.0 μm and 0.4–25.0 μm [41, 42]. The calculated NLO coefficients are $d_{15} = 9.21$, $d_{22} = 9.22$, $d_{33} = 6.91$ pm/V for $(SbI_3).(S_8)_3$ and $d_{15} = 3.40$, $d_{22} = 6.21$, $d_{33} = 0.73$ pm/V for $(AsI_3).(S_8)_3$. The chalcohalides (SnI_4). $(S_8)_2$ has space group Fdd_2 [44] and a wide IR window in the range of 2.5–25.0 μm. The energy gap of (SnI_4). $(S_8)_2$ is 2.17 eV. The SHG response of (SnI_4). $(S_8)_2$ is 0.5 times stronger than that of $AgGaS_2$ at 2.1 μm.

4.3 Lewis acid adduct chalcohalides

Lewis acid adduct chalcohalides have sulfur-nitrogen rings with a variety of structures. These chalcohalides have moderate band gaps and IR-NLO properties (**Table 5**). A new chalcohalide $(NSF)_4$ [45] has a space group of P-421c. The calculated energy gaps are 4.57 eV (HSE06 method) and 3.58 eV (GGA method). For a high LDT, the large bandgap is more advantageous. The birefringence for $(NSF)_4$ is 0.220 at 1064 nm and the calculated NLO coefficient for $(NSF)_4$ is $d_{14} = 3.20$ pm/V. Increased bandgap occurs due to the large electronegativity of F atoms in the

Crystal	Space group	Band gap (HSE06) (ev)	NLO coefficient (pm/V)
$(NSF)_4$	P-421c	4.57	d_{14} = 3.20
$S_3N_5PF_2$	R3m	3.49	d_{15} = 1.71, d_{21} = 0.19 and d_{33} = 3.69

Table 5.
Space group and band gap, and value of Lewis acid adduct chalcohalides.

compound. $S_3N_5PF_2$ crystal [46] has an R3m space group. It has a wide bandgap of 3.49 eV (HSE06). The estimated birefringence is around 0.110 at 1064 nm and NLO coefficients are d_{15} = 1.71, d_{21} = 0.19 and d_{33} = 3.69 pm/V. The d_{33} coefficient is around 9.2 times stronger than that of KDP.

4.4 Main group element clusters chalcohalides

The chalcohalides $(Bi_4S_4)(AlCl_4)_4$ belong to the space group I-4 (**Figure 4**) [47] and it has two classes of main group component clusters that are in $AlCl_4$ tetrahedron and Bi_4S_4 cube. The energy gap of $(Bi_4S_4)(AlCl_4)_4$ is calculated to be 3.59 eV. $(Bi_4S_4)(AlCl_4)_4$ is the largest main group element cluster chalcohalide, which has an NLO coefficient value of d_{14} = 1.52 pm/V and it is 3.8 times stronger than that of KDP. The DFT method showed that the main group component cluster chalcohalides has insignificant NLO coefficients.

4.5 Other main groups of metal chalcohalides

Both In_5S_5Cl [48] and In_5S_5Br [49] have isostructural properties, which are having the space groups Pmn21. The In_5S_5Cl and In_5S_5Br materials have band gap of 1.76 and 1.84 eV, respectively (**Table 6**). The estimated NLO coefficients are found to be d_{15} = 0.36, d_{24} = 2.83, d_{33} = 13.38 pm/V for In_5S_5Cl and d_{15} = 2.07, d_{24} = 2.21, d_{33} = 7.38 pm/V for In_5S_5Br. When compared with the NLO coefficients of AGS, the main group metal chalcohalides In_5S_5Cl and In_5S_5Br have d_{33}=1.0 and 0.5 times than that of AGS. The $(CS_3N_2Br)Br_3$ main group metal chalcohalides [50] crystallizes in space group Pna_{21}. The estimated energy gap for $(CS_3N_2Br)Br_3$ is 2.21 eV ($HSE0_6$), and the NLO coefficients of $(CS_3N_2Br)Br_3$ are d_{15} = 8.90, d_{24} = 5.40, d_{33} = 1.00 pm/V. From the NLO coefficient result, d_{15} has a good NLO coefficient, which is about 0.6 times that of AGS.

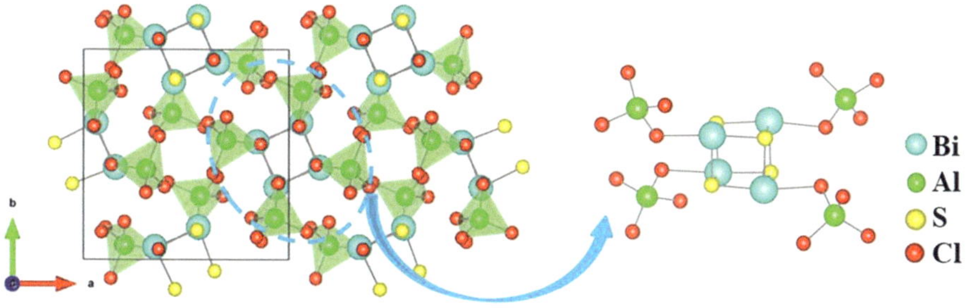

Figure 4.
Crystal structure of $(Bi_4S_4)(AlCl_4)_4$.

Crystal	Space group	Band gap (HSE06) ev	NLO coefficient (pm/V)	NLO coefficient (AGS) times
In_5S_5Cl	Pmn21	1.76	d_{15} = 0.36, d_{24} = 2.83, d_{33} = 13.38	1.0
In_5S_5Br	Pmn21	1.84	d_{15} = 2.07, d_{24} = 2.21, d_{33} = 7.38	0.5
$(CS_3N_2Br)Br_3$	Pna_{21}	2.21	d_{15} = 8.90, d_{24} = 5.40, d_{33} = 1.00	0.6

Table 6.
Space group, band gap, and SHG value of other main groups of metal chalcohalides.

4.6 Transition metal chalcohalides

Asymmetric distribution of electron clouds is generally caused by polyhedron with the d_{10} transition metals because of the dp orbital hybridization and group distortion, so it produces high SHG behavior. Due to the dp hybridization, the compound has a red-shifted absorption edge, this might reduce the bandgap of IR-NLO materials. To enhance the chalcogenides energy gap, halogen elements are introduced, that is, by combining cations with d10 configuration, which led to the equilibrium among energy gaps and SHG behavior in the IR-NLO crystals. So, they separated the asymmetric chalcohalides with a d_{10} electronic configuration. The transition metal chalcohalide Ag_2HgSI_2 has the space group Cmc_{21} [51]. The bandgap of Ag_2HgSI_2 is 2.65 eV and it is compared with AGS of (2.70 eV). The SHG of Ag_2HgSI_2 is 4.2 times stronger than that of KDP. The theoretical birefringence of Ag_2HgSI_2 is 0.210 at 1064 nm. The crystal $(P4S_3)_3(CuCl)_7$ belongs to the space group $P_{31}c$ [52]. The bandgap of $(P_4S_3)_3(CuCl)_7$ is 2.77 eV ($HSE0_6$ method) and NLO coefficients of $(P_4S_3)_3(CuCl)_7$ are d_{33} = 3.34 and d_{15} = d_{24} = 1.81 pm/V. In the NLO coefficients, the d_{33} value is about 8.5 times that of KDP and has a birefringence value of 0.150 at 1064 nm.

5. Oxide crystal

Normally, oxide-based materials have a high laser damage threshold and many of them have limited IR transparency within the range of 6 μm. Here, we would like to explain the new oxide-based NLO materials, which are very important for the development of IR transparent regions beyond 6 μm. Several new NLO oxide crystals were identified, which have high IR cutoff wavelengths that are up to 6 μm. For instance β-$BaTeMo_2O_9$ [53], $MnTeMoO_6$ [54], $Cs_2TeW_3O_{12}$ [55], $V_2Te_2O_9$ [56], Li_3VO_4 [57] , M_2LiVO_4 (M = Rb, Cs) [58], $Li_2K_4TiOGe_4O_{12}$ [59], and $Rb_4Li_2TiOGe_4O_{12}$ [60]. Moreover, for the oxide-based NLO material, IR transparent range is a difficult factor to attain. $Pb_{17}O_8Cl_{18}$ was new IR-NLO material discovered by Pan and Poeppelmeier et al. [61] in 2015. $Pb_{17}O_8Cl_{18}$ single crystal was synthesized by modified spontaneous crystallization in an open model. Tao's et al. [62] reported a new $LiNbO_3$-type NLO crystal. Li_2ZrTeO_6 crystal (size: 16 × 15 × 12 mm^3) is grown by top-seeded solution growth (TSSG) method. To maintain the structural qualities of $LiNbO_3$, $Zr^{4+}$and Te^{6+} were substituted for Nb^{5+} to form Li_2ZrTeO_6 and it crystallized in the trigonal crystal system with space group *R*3 (**Figure 5**). $LiNbO_3$ belongs to the space group, R_3c, which has a close structural feature with Li_2ZrTeO_6 [63]. $La_3SnGa_5O_{14}$ is a new

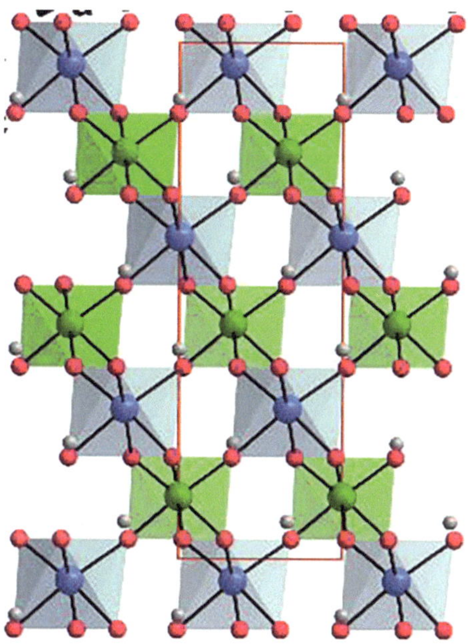

Figure 5.
Crystal structures of Li_2ZrTeO_6.

IR-NLO crystal, it belongs to the langasite family and has a space group *P*321 [64]. Polycrystalline $La_3SnGa_5O_{14}$ was produced using a solid-state reaction and a single crystal were grown by Czochralski method. The single crystal of $Pb_{17}O_8Cl_{18}$ has a bandgap of 3.44 eV and it has high IR transparency (13.9 μm). $La_3SnGa_5O_{14}$ has a wide energy gap value of 4.60 eV and transparency of 10 μm.

The SHG responses of $Pb_{17}O_8Cl_{18}$ showed a response at 2090 nm and 1064 nm, which is phase-matchable, and it is two times stronger than that of $AgGaS_2$ and four times higher than that of KDP. Li_2ZrTeO_6 showed a huge powder SHG behavior at 1064 nm, which is 2.5 times higher than that of KDP [63]. For Li_2ZrTeO_6 and $LiNbO_3$, there is a variation in the SHG responses, which is closely correlated, and it is due to the different sizes of octahedral distortions in their crystal structure. The SHG efficiency of $La_3SnGa_5O_{14}$ is 0.4 times that of $AgGaS_2$ and when compared to $AgGaS_2$ (13 μm) it has wide IR transparency. $Pb_{17}O_8Cl_{18}$ has an LDT of 12.8 MW/cm^2 (**Table 7**). These characteristics reveal that $Pb_{17}O_8Cl_{18}$ is one of the good mid-IR-NLO crystals for the next generation. Li_2ZrTeO_6 has an outstanding optical performance and a very high LDT greater than 1.3 GW/cm^2

Crystal	Space group	Bandgap (eV)	IR range	SHG	LDT (× AGS, MW/cm2)
Li_2ZrTeO_6	*R*3	4.06	7.4 μm	2.5 times (KDP)	> 1300 MW/cm^2
$La_3SnGa_5O_{14}$	*P*321	4.60 eV	10 μm	0.4 times that of $AgGaS_2$	28 × AGS, 846 MW/cm2
$Pb_{17}O_8Cl_{18}$	Fmm2	3.44 eV	13.9 μm	2 times ($AgGaS_2$) and 4 times (KDP)	12.8 × AGS (on powder)

Table 7.
Space group, band gap, transparency range, and SHG value of oxide crystal.

and it was more than 22 times that of $LiNbO_3$. It has a higher IR transparent range that is up to 7.4 µm. $La_3SnGa_5O_{14}$,which has the highest LDT of 846 MW/cm^2, is an alternative NLO crystal in the mid-IR region. It is transparent beyond 10 µm. The langasite family provided valuable information to design a new NLO crystal in the IR area.

6. Halides

A halide has a binary form in which one region is a halogen and the other region is a component or radical that is less electronegative/electropositive than the halogen [65]. For preparing NLO halides, the physical, chemical, and crystallographic aspects are important. In halides, hyperpolarizabilities were used to find out the NLO susceptibilities. Moreover, these crystals have high LDT and good mechanical properties. $HgBr_2$ crystal was grown using a lowering temperature technique with a size of about 15 × 15 × 1.5 mm^3 [66]. Tl_3PbBr_5, Tl_4PbI_6, Tl_4HgI_6, and Tl_3PbI_5 were grown by the vertical Bridgman method [5]. The 3:1:5 ratio showed orthorhombic symmetry and the 4:1:6 ratio indicated tetragonal symmetry. In this method, they maintained the temperature gradients in the range from 20 K/cm to 30 K/cm and cooling rates were in the range from 5 to 10 K/h, which showed a growth of 1 cm/day and 3–5 cm/day. Tl_3PbCl_5 [67] and Tl_3PbBr_5 [68] single crystals were grown by Bridgman-Stockbarger technique. Tl_3PbI_5 is colorless and Tl_3PbBr_5 (**Figure 6**) crystal is a yellow color and transparent. Tl_4HgI_6 crystal is grown using Bridgman-Stockbarger method [69, 70] and they melt consistently at 396°C. The crystal was red and when the iodine concentration increases in the stoichiometric ratio, the crystal becomes changed to black. It belongs to the point symmetry group C_4v. $BaMgF_4$ crystal was grown using the Czochralski technique [71]. It belongs to the pyroelectric fluoride group $BaMF_4$ (M = Mg, Co, Ni, Zn) and it has space group Cmc_{21}. $SrAlF_5$ crystal belongs to the class of uniaxial ferroelectric [72] and is grown using Czochralski technique [71]. Tl_3PbCl_5 and Tl_3PbBr_5 are nonhygroscopic. In Tl_3PbBr_5, the phase transition is observed at ~237°C, and for Tl_3PbCl_5, phase transition is noted at 171°C. Tl_2HgI_4 compound has a melting point temperature of 318°C.

$HgBr_2$ crystal showed good phase matchable SHG efficiency, which is 10 times greater than that of KDP and has transparency between 2.5 and 25 mm. It covers the whole mid-IR range. Tl_3PbCl_5 and Tl_3PbBr_5 have transparency of 0.5–20 mm and 0.65–24 mm, respectively. The Tl_4HgI_6 crystal is optically positive and transparent vfrom 1.2 to 40 mm. $BaMgF_4$ is a ferroelectric fluoride due to its wide transparency between 125 nm and 13 mm, and it can be used for UV and mid-IR optical applications [71, 72]. In the UV region, the shortest band is noted at 368 nm, which represents the potential behavior of $BaMgF_4$ as a nonlinear material. It can be used for the production of all solid-state lasers and mid-IR wavelength areas. $BaMgF_4$ and $SrAlF_5$ crystals are promising crystals for solid-state lasers. The LDT value of the $HgBr_2$ crystal is 0.3 GW/cm^2.

Figure 6.
Crystal of Tl_3PbBr_5.

7. Oxyhalides

To develop high efficiency in mid-IR-NLO crystals, it is important to have an oxide-based system with an expandable IR transparent range. Many NLO oxides have a range from 3 to 5 μm atmospheric windows. When compared with the $AgGaS_2$ (13 μm), the oxide-based materials, such as $La_3SnGa_5O_{14}$ and $Pb_{17}O_8Cl_{18,}$ have wide IR transparency between 12 and 13.9 μm. In this lead oxyhalide, NLO crystal $Pb_{17}O_8Cl_{18}$ plays an outstanding overall property. To obtain high LDT, halide and oxide-based crystals with high bandgaps are used. Nowadays, the mixture of heavy metal lone pair cation, Pb^{2+}, and mixed oxyhalides are focused on IR applications. Xinglong Chen et al. discovered the lead mixed oxyhalides, such as $Pb_{13}O_6C_{19}Br_5$, $Pb_{13}O_6C_{17}Br_{7,}$ and $Pb_{13}O_6C_{14}Br_{10}$ [73], which have broad IR transparency up to 14 μm, high SHG behavior (0.6–0.9 × $AgGaS_2$) and wide bandgaps from 3.05 to 3.21 eV. $Pb_{13}O_6C_{19}Br_5$ single crystal has the size of 2.9 × 1.3 × 0.5 cm^3, which was grown using the top-seeded solution growth (TSSG) method. It has a wide transparent range from 0.384 to 14.0 *μ*m and a high LDT value (14.6 × $AgGaS_2$). Many crystals, such as $APbCO_3F$ (A = Rb, Cs), Pb_2BO_3Cl, $Cs_3VO(O_2)_2CO_3$, $Bi_3TeBO_{9,}$ and $BiFSeO_3$ [74–78], have been reported with outstanding properties and it is a very good material for visible/near-IR non-linear optical applications. $Pb_{13}O_6C_{14}Br_{10}$, $Pb_{13}O_6C_{17}Br_7$, and $Pb_{13}O_6C_{19}Br_5$ have orthorhombic crystal structures with space group *Fmm*2 (**Table 8**). These three crystals are isomorphic.

By using Czochralski and flux method, oxide-based crystals can be obtained. The $Pb_{13}O_6C_{14}Br_{10}$, $Pb_{13}O_6C_{17}Br_7$, and $Pb_{13}O_6C_{19}Br_5$ crystals were grown by flux method and the self-flux method was adopted for $PbCl_2$-$PbBr_2$. The crystals obtained by this method were optically transparent and they showed a good growth rate. $Pb_{13}O_6Cl_9Br_5$ was used to grow large-size crystals using the TSSG technique. After many technical optimizations, two big size crystals (Dimensions up to 2.9 × 1.3 × 0.5 and 3.7 × 0.4 × 0.7 cm^3) were grown using the [001] and [100] oriented seeds. They have good transparency and good growth speed. Therefore, to get a better quality crystal the growth parameters (cooling rate, rotation speed, and temperature gradient) are very important in the crystallization process. DSC curves of $Pb_{13}O_6Cl_4Br_{10}$, $Pb_{13}O_6Cl_7Br_{7,}$ and $Pb_{13}O_6Cl_9Br_5$ show that each of them has one endothermic peak at 501°C, 504°C, and 508°C, respectively, which belongs to the melting point and these crystals have two exothermic peaks, which indicate the decomposing of the compounds. Due to the volatility of the halide materials, there is no weight loss before 490°C in the TGA curves. From the DSC and TGA, it was concluded that these compounds have high thermal stability up to 490°C. $Pb_{13}O_6C_{19}Br_5$, $Pb_{13}O_6Cl_7Br_{7,}$ and $Pb_{13}O_6C_{14}Br_{10}$ have high reflectance wavelengths in the region between 500 and 2500 nm. When compared with $AgGaS_2$ (2.67 eV, 0.53–13 μm) and $ZnGeP_2$ (1.68 eV, 0.74–12 μm), the crystals of $Pb_{13}O_6C_{14}Br_{10}$, $Pb_{13}O_6C_{17}Br_7$, and $Pb_{13}O_6C_{19}Br_5$ own greater energy gaps and good transparency in IR region [79, 80].

Crystal	Space group	SHG ($AgGaS_2$)	LDT ($AgGaS_2$)
$Pb_{13}O_6Cl_4Br_{10}$	*Fmm*2	0.6	3.0
$Pb_{13}O_6Cl_7Br_7$	*Fmm*2	0.8	3.2
$Pb_{13}O_6Cl_9Br_5$	*Fmm*2	0.9	4.0

Table 8.
Space group and SHG value of oxyhalides.

In the application of high-power laser systems, the NLO crystal with laser-induced damage is one of the biggest problems. The crystals which have wider bandgaps are subjected to higher LDTs. Polycrystalline samples are used for the laser-induced damage threshold evaluation, and the polycrystalline material of $AgGaS_2$ was used as reference material. The $Pb_{13}O_6C_{14}Br_{10}$, $Pb_{13}O_6C_{17}Br_7$, and $Pb_{13}O_6C_{19}Br_5$ have large LDTs values, which are 3.0, 3.2, and 4.0 times higher than that of $AgGaS_2$. The $Pb_{13}O_6C_{19}Br_5$ crystal has LDT of 439 MW/cm^2, this is 14.6 times higher than that of the $AgGaS_2$ crystal (30 MW/cm^2). The second harmonic generation intensity is 0.5 times higher than that of $AgGaS_2$. The SHG value of $Pb_{13}O_6C_{14}Br_{10}$, $Pb_{13}O_6C_{17}Br_7$, and $Pb_{13}O_6C_{19}Br_5$ are found to be 0.6, 0.8, and 0.9, respectively. From this, we concluded that SHG responses of each material have all phase-matchable when it is under the 2090 nm wavelength. Due to the increase in particle size, it showed a positive movement in SHG response.

8. Conclusion

In summary, significant progress has been attained in the search for new, favorable IR nonlinear crystals, such as chalcogenides, chalcohalides, oxides, halides, and oxyhalides, for producing coherent energy in the MIR. For a good NLO crystal, the crystals must have good physical and chemical properties, such as wide transparency range, LDT, chemical stability, birefringence, and nonlinear susceptibility. Nevertheless, for practical applications, the crystals should have high qualities with promising properties. For promising optical applications, accurate optical properties should be measured. For MIR applications, high bandgap value, LDT, and NLO susceptibility optical transparency are important. Various IR crystals fulfill the basic conditions of IR-NLO applications. The significant behavior of these crystals recommends that this crystal can be used in different optical applications.

DOI: http://dx.doi.org/10.5772/intechopen.108173

Author details

Senthil Kumar Chandran[1]*, Chinnakannu Elavarasi[2], Srinivasan Manikam[3] and John James Gnanapragasam[4]

1 Department of Physics, Government Arts and Science College, Hosur, Tamil Nadu, India

2 Sri Vijay Vidyalaya College of Arts and Science, Dharmapuri, Tamil Nadu, India

3 Department of Physics, Center for Crystal Growth, SSN College of Engineering, Kalavakkam, Tamil Nadu, India

4 Department of Physics, Government Arts College, Trichy, Tamil Nadu, India

*Address all correspondence to: senthilkumarchandran89@gmail.com

References

[1] Yan M, Xue HG, Guo SP. Recent achievements in lone-pair cation-based infrared second-order nonlinear optical materials. Crystal Growth & Design. 2020;**21**(1):698-720

[2] Chen X, Ok KM. Recent advances in oxide-based nonlinear optical materials with wide infrared transparency beyond 6 μm. Chemistry, an Asian Journal. 2020;**15**(22):3709-3716

[3] Liang F, Kang L, Lin Z, Wu Y. Mid-infrared nonlinear optical materials based on metal chalcogenides: Structure–property relationship. Crystal Growth & Design. 2017;**17**(4):2254-2289

[4] Gao L, Huang J, Guo S, Yang Z, Pan S. Structure-property survey and computer-assisted screening of mid-infrared nonlinear optical chalcohalides. Coordination Chemistry Reviews. 2020;**421**:213379

[5] Jiang XM, Guo SP, Zeng HY, Zhang MJ, Guo GC. Large crystal growth and new crystal exploration of mid-infrared second-order nonlinear optical materials. Structure-Property Relationships in Non-Linear Optical Crystals II. 2012;**2012**:1-43

[6] Isaenko LI, Yelisseyev AP. Recent studies of nonlinear chalcogenide crystals for the mid-IR. Semiconductor Science and Technology. 2016;**31**(12):123001

[7] Shi ZH, Chi Y, Sun ZD, Liu W, Guo SP. Sn2Ga2S5: A type of IR nonlinear-optical material. Inorganic Chemistry. 2019;**58**(18):12002

[8] Nguyen V, Ji B, Wu K, Zhang B, Wang J. Unprecedented mid-infrared nonlinear optical materials achieved by crystal structure engineering, a case study of (KX) P 2 S 6 (X= Sb, Bi, Ba). Chemical Science. 2022;**13**(9):2640-2648

[9] Wu C, Lin L, Jiang X, Lin Z, Huang Z, Humphrey MG, et al. K5 (W3O9F4) (IO3): An efficient mid-infrared nonlinear optical compound with high laser damage threshold. Chemistry of Materials. 2019;**31**(24):10100-10108

[10] Yu T, Wang S, Zhang X, Li C, Qiao J, Jia N, et al. MnSiP2: A new Mid-IR ternary phosphide with strong SHG effect and ultrabroad transparency range. Chemistry of Materials. 2019;**31**(6):2010-2018

[11] Guo SP, Chi Y, Guo GC. Recent achievements on middle and far-infrared second-order nonlinear optical materials. Coordination Chemistry Reviews. 2017;**335**:44-57

[12] Guo Y, Liang F, Yin W, Li Z, Luo X, Lin ZS, et al. BaHgGeSe4 and SrHgGeSe4: Two new Hg-based infrared nonlinear optical materials. Chemistry of Materials. 2019;**31**(8):3034-3040

[13] Li MY, Ma Z, Li B, Wu XT, Lin H, Zhu QL. HgCuPS4: An exceptional infrared nonlinear optical material with defect diamond-like structure. Chemistry of Materials. 2020;**32**(10):4331-4339

[14] Kushwaha AK, Kalita H, Suman S, Bhardwaj A, Ghosh R. Two-dimensional (2D) thermoelectric materials. Thermoelectricity and Advanced Thermoelectric Materials. 2021:233-260

[15] Muslih EY, Munir B, Khan MM. Advances in chalcogenides and chalcogenides-based nanomaterials such as sulfides, selenides, and tellurides. Chalcogenide-Based Nanomaterials as Photocatalysts. 2021:7-31

[16] Shi W, Ding YJ, Fernelius N, Vodopyanov K. Efficient, tunable, and coherent 0.18-5.27-THz source based on GaSe crystal. Optics Letters. 2002;**27**(16):1454-1456

[17] Abudurusuli A, Li J, Pan S. A review on the recently developed promising infrared nonlinear optical materials. Dalton Transactions. 2021;**50**(9):3155-3160

[18] Nitsche R, Bölsterli HU, Lichtensteiger M. Crystal growth by chemical transport reactions—I: Binary, ternary, and mixed-crystal chalcogenides. Journal of Physics and Chemistry of Solids. 1961;**21**:199-205

[19] Bajpai PK, Yadav S, Tiwari A, Virk HS. Recent advances in the synthesis and characterization of chalcogenide nanoparticles. Solid State Phenomena. 2015;**222**:187-233

[20] Jiang X, Kang L, Luo S, Gong P, Lee MH, Lin Z. Development of nonlinear optical materials promoted by density functional theory simulations. International Journal of Modern Physics B. 2014;**28**(27):1430018

[21] Ohmer MC, Goldstein JT, Zelmon DE, Saxler AW, Hegde SM, Wolf JD, et al. Infrared properties of AgGaTe 2, a nonlinear optical chalcopyrite semiconductor. Journal of Applied Physics. 1999;**86**(1):94-99

[22] Schunemann PG, Setzler SD, Pollak TM, Ohmer MC, Goldstein JT, Zelmon DE. Crystal growth and properties of AgGaTe2. Journal of Crystal Growth. 2000;**211**:242-246

[23] Isaenko L, Yelisseyev A, Lobanov S, Titov A, Petrov V, Zondy JJ, et al. Growth and properties of LiGaX2 (X= S, Se, Te) single crystals for nonlinear optical applications in the mid-IR. Crystal Research and Technology: Journal of Experimental and Industrial Crystallography. 2003;**38**:379-387

[24] Yelisseyev A, Isaenko L, Lovanov S, Zondy JJ. Advanced solid state lasers. OSA Trends in Optics and Photonics Series. 2000;**34**:561-569

[25] Zondy JJ, Bielsa F, Douillet A, Hilico L, Acef O, Petrov V, et al. Frequency doubling of CO 2 laser radiation at 10.6 μm in the highly nonlinear chalcopyrite LiGaTe 2. Optics Letters. 2007;**32**(12):1722-1724

[26] Yao J, Mei D, Bai L, Lin Z, Yin W, Fu P, et al. BaGa4Se7: A new congruent-melting IR nonlinear optical material. Inorganic Chemistry. 2010;**49**(20):9212-9216

[27] Yao J, Yin W, Feng K, Li X, Mei D, Lu Q, et al. Growth and characterization of BaGa4Se7 crystal. Journal of Crystal Growth. 2012;**346**(1):1-4

[28] Isaenko L, Yelisseyev A, Lobanov S, Krinitsin P, Petrov V, Zondy JJ. Ternary chalcogenides LiBC2 (B= In, Ga; C= S, Se, Te) for mid-IR nonlinear optics. Journal of Non-Crystalline Solids. 2006;**352**:2439-2443

[29] Yelisseyev AP, Drebushchak VA, Titov AS, Isaenko LI, Lobanov SI, Lyapunov KM, et al. Thermal properties of the mid-infrared nonlinear crystal LiInSe 2. Journal of Applied Physics. 2004;**96**(7):3659-3665

[30] Petrov V, Yelisseyev A, Isaenko L, Lobanov S, Titov A, Zondy JJ. Second-harmonic generation and optical parametric amplification in the mid-IR with orthorhombic biaxial crystals LiGaS2 and LiGaSe2. Applied Physics B. 2004;**78**(5):543-546

[31] Badikov V, Badikov D, Shevyrdyaeva G, Tyazhev A, Marchev G,

Panyutin V, et al. BaGa 4 S 7: Wide-bandgap phase-matchable nonlinear crystal for the mid-infrared. Optical Materials Express. 2011;**1**(3):316-320

[32] Tyazhev A, Kolker D, Marchev G, Badikov V, Badikov D, Shevyrdyaeva G, et al. Midinfrared optical parametric oscillator based on the wide-bandgap BaGa 4 S 7 nonlinear crystal. Optics Letters. 2012;**37**(19):4146-4148

[33] Medina-Ramírez I, Hernández-Ramírez A, Maya-Trevino ML. Synthesis methods for photocatalytic materials. In Photocatalytic Semiconductors. 2015:69-102

[34] Isaenko LI, Yelisseyev AP, Lobanov SI, Krinitsin PG, Molokeev MS. Structure and optical properties of Li2Ga2GeS6 nonlinear crystal. Optical Materials. 2015;**47**:413-419

[35] Yelisseyev AP, Isaenko LI, Krinitsin P, Liang F, Goloshumova AA, Naumov DY, et al. Crystal growth, structure, and optical properties of LiGaGe2Se6. Inorganic Chemistry. 2016;**55**(17):8672-8680

[36] Yin W, Feng K, Hao W, Yao J, Wu Y. Synthesis, structure, and properties of li2in2mq6 (m= si, ge; q= s, se): A new series of ir nonlinear optical materials. Inorganic Chemistry. 2012;**51**(10):5839-5843

[37] Lin X, Guo Y, Ye N. BaGa2GeX6 (X= S, Se): New mid-IR nonlinear optical crystals with large band gaps. Journal of Solid State Chemistry. 2012;**195**:172-177

[38] Liu PF, Li YY, Zheng YJ, Yu JS, Duan RH, Chen H, et al. Tailored synthesis of nonlinear optical quaternary chalcohalides: Ba 4 Ge 3 S 9 Cl 2, Ba 4 Si 3 Se 9 Cl 2 and Ba 4 Ge 3 Se 9 Cl 2. Dalton Transactions. 2017;**46**(8):2715-2721

[39] Feng K, Kang L, Lin Z, Yao J, Wu Y. Noncentrosymmetric chalcohalide NaBa 4 Ge 3 S 10 Cl with large bandgap and IR NLO response. Journal of Materials Chemistry C. 2014;**2**(23):4590-4596

[40] Liu BW, Zeng HY, Jiang XM, Wang GE, Li SF, Xu L, et al. [A 3 X][Ga 3 PS 8](A= K, Rb; X= Cl, Br): Promising IR non-linear optical materials exhibiting concurrently strong second-harmonic generation and high laser-induced damage thresholds. Chemical Science. 2016;**7**(9):6273-6277

[41] Guo SP, Sun ZD, Chi Y, Xue HG. Adduct-type IR nonlinear-optical crystal SbI3·(S8) 3 with a large second-harmonic generation and a high laser-induced damage threshold. Inorganic Chemistry. 2018;**57**(17):11282-11288

[42] Samoc A, Samoc M, Prasad PN, Krajewska-Cizio A. Second-harmonic generation in the crystalline complex antimony triiodide–sulfur. JOSA B. 1992;**9**(10):1819-1824

[43] Lu ZT, Sun ZD, Chi Y, Xue HG, Guo SP. Balanced second-order nonlinear optical properties of adducts CHI3·(S8) 3 and AsI3·(S8) 3: A systematic survey. Inorganic Chemistry. 2019;**58**(7):4619-4625

[44] Guo SP, Chi Y, Xue HG. SnI4·(S8) 2: A novel adduct-type infrared second-order nonlinear optical crystal. Angewandte Chemie. 2018;**130**(36):11714-11717

[45] Wiegers GT, Vos A. Refinement of the structure of tetrathiazylfluoride (NSF)4. Acta Crystallographica. 1963;**16**(2):152-153

[46] Weiss J, Ruppert I, Appel R. Die Kristall-und Molekülstruktur von S3N5PF2. Zeitschrift für Anorganische

und Allgemeine Chemie. 1974;**406**:329-336

[47] Beck J, Schlueter S, Zotov N. The cube-shaped main group element clusters (Bi4S4)4+ and (Bi4Se4)4+-synthesis from chloroaluminate melts, crystal structures and vibrational spectra. Zeitschrift für Anorganische und Allgemeine Chemie. 2004;**630**:2512-2519

[48] Nickel V, Deiseroth HJ, Kienle L, Duppel V, Reiner C. Polymorphism of In5S5Cl–X-ray and HRTEM-Investigations. Allgemeine Chemie. 2010;**636**:79-84

[49] Deiseroth HJ, Reiner C, Xhaxhiu K, Schlosser M, Kienle L. X-Ray and transmission electron microscopy investigations of the new solids In5S5Cl, In5Se5Cl, In5S5Br, and In5Se5Br. Zeitschrift für Anorganische und Allgemeine Chemie. 2004;**630**:2319-2328

[50] Wolmershäuser G, Krüger C, Tsay YH. [CS3N2Br]+ Br–ein Produkt der Bromierung von S4N4 in Schwefelkohlenstoff. Chemische Berichte. 1982;**115**(3):1126-1131

[51] Keller HL, Wimbert L. Über Münzmetall-Quecksilber-Chalkogenidhalogenide. III Zur Kristallstruktur von Ag2HgSI2. Zeitschrift für Anorganische und Allgemeine Chemie. 2003;**629**(12-13):2337-2340

[52] Biegerl A, Brunner E, Gröger C, Scheer M, Wachter J, Zabel M. The unexpected versatility of P4S3 as a building block in polymeric copper halide networks: 2, 3-P, 1, 2, 3-P and all-P coordination. Chemistry – A European Journal. 2007;**13**(33):9270-9276

[53] Zhang W, Tao X, Zhang C, Gao Z, Zhang Y, Yu W, et al. Bulk growth and characterization of a novel nonlinear optical crystal BaTeMo2O9. Crystal Growth & Design. 2008;**8**(1):304-307

[54] Jin CG, Li Z, Huang LX, He MZ. Top-seeded solution growth and characterization of a novel nonlinear optical crystal MnTeMoO6. Journal of Crystal Growth. 2013;**369**:43-46

[55] Zhang J, Tao X, Sun Y, Zhang Z, Zhang C, Gao Z, et al. Top-seeded solution growth, morphology, and properties of a polar crystal Cs2TeMo3O12. Crystal Growth & Design. 2011;**11**(5):1863-1868

[56] Zhang W, Wang X, Shen G, Shen D. Top-seeded growth, optical properties and theoretical studies of noncentrosymmetric Te2V2O9. Crystal Research and Technology. 2012;**47**:163-168

[57] Chen Z, Zhang Z, Dong X, Shi Y, Liu Y, Jing Q. Li3VO4: A promising mid-infrared nonlinear optical material with large laser damage threshold. Crystal Growth & Design. 2017;**17**:2792-2800

[58] Han G, Wang Y, Su X, Yang Z, Pan S. Growth, properties, and theoretical analysis of M2LiVO4 (M = Rb, Cs) crystals: Two potential mid-infrared nonlinear optical materials. Scientific Reports. 2017;**7**:1901

[59] Xu J, Wu H, Yu H, Zhang W, Hu Z, Wang J, et al. Li2K4TiOGe4O12: A stable mid-infrared nonlinear optical material. Chemistry of Materials. 2020;**32**:906-912

[60] Xi M, Tang C, Li R. Rb4Li2TiOGe4O12: A titanyl nonlinear optical material with the widest transparency range. Angewandte Chemie, International Edition. 2019;**58**:18257-18260

[61] Zhang H, Zhang M, Pan S, Dong X, Yang Z, Hou X, et al. Pb17O8Cl18: A

promising IR nonlinear optical material with large laser damage threshold synthesized in an open system. Journal of the American Chemical Society. 2015;**137**:8360-8363

[62] Lu W, Gao Z, Liu X, Tian X, Wu Q, Li C, et al. Rational design of a LiNbO3-like nonlinear optical crystal, Li2ZrTeO6, with high laser-damage threshold and wide Mid-IR transparency window. Journal of the American Chemical Society. 2018;**140**:13089-13096. DOI: 10.1021/jacs.8b08803

[63] Ok KM, Halasyamani PS, Casanova D, Llunell M, Alemany P, Alvarez S. Distortions in octahedrally coordinated d0 transition metal oxides: A continuous symmetry measures approach. Chemistry of Materials. 2006;**18**:3176-3183

[64] Lan H, Liang F, Jiang X, Zhang C, Yu H, Lin Z, et al. Pushing nonlinear optical oxides into the mid-infrared spectral region beyond 10 μm: Design, synthesis, and characterization of La3SnGa5O14. Journal of the American Chemical Society. 2018;**140**:4684-4690

[65] Hagemann M, Weber HJ. Are ternary halides useful materials for nonlinear optical applications? Applied Physics A. 1996;**63**(1):67-74

[66] Zhu T, Chen X, Qin J. Research progress on mid-IR nonlinear optical crystals with high laser damage threshold in China. Frontiers of Chemistry in China. 2011;**6**(1):1-8

[67] Singh NB, Suhre DR, Green K, Fernelius N, Hopkins FK. Periodically poled materials for long-wavelength infrared (LWIR) NLO applications. Journal of Crystal Growth. 2005;**274**:132-137

[68] Ferrier A, Velázquez M, Pérez O, Grebille D, Portier X, Moncorgé R. Crystal growth and characterization of the non-centrosymmetric compound Tl3PbCl5. Journal of Crystal Growth. 2006;**291**(2):375-384

[69] Ferrier A, Velázquez M, Portier X, Doualan JL, Moncorgé R. Tl3PbBr5: A possible crystal candidate for middle infrared nonlinear optics. Journal of Crystal Growth. 2006;**289**(1):357-365

[70] Badikov DV, Badikov VV, Kuz'micheva GM, Panyutin VL, Rybakov VB, Chizhikov VI, et al. Growth and X-ray diffraction study of Tl4HgI6 crystals. InorganicMaterials. 2004;**40**:314-320

[71] Shimamura K, Víllora EG, Muramatsu K, Ichinose N. Advantageous growth characteristics and properties of SrAlF5 compared with BaMgF4 for UV/VUV nonlinear optical applications. Journal of Crystal Growth. 2005;**275**:128-134

[72] Ravez J, Simon A, Chaminade JP. Ferroelectric behavior and phase transition at 715 K in $SrAIF_5$. Journal of Applied Physics. 1981;**52**:4740-4743

[73] Chen X, Jo H, Ok KM. Lead mixed oxyhalides satisfying all fundamental requirements for high-performance mid-infrared nonlinear optical materials. Angewandte Chemie. 2020;**132**(19):7584-7590

[74] Zou G, Lin C, Jo H, Nam G, You TS, Ok KM. Pb2BO3Cl: A Tailor-Made polar lead borate chloride with very strong second harmonic generation. Angewandte Chemie. 2016;**128**(39):12257-12261

[75] Zou G, Lin Z, Zeng H, Jo H, Lim SJ, You TS, et al. Cs 3 VO (O 2) 2 CO 3: An exceptionally thermostable carbonatoperoxovanadate with an extremely large second-harmonic

generation response. Chemical Science. 2018;**9**(48):8957-8961

[76] Xia M, Jiang X, Lin Z, Li R. "All-three-in-one": A new bismuth–tellurium–borate Bi3TeBO9 exhibiting strong second harmonic generation response. Journal of the American Chemical Society. 2016;**138**(43):14190-14193

[77] Zou G, Ye N, Huang L, Lin X. Alkaline-alkaline earth fluoride carbonate crystals ABCO3F (A= K, Rb, Cs; B= Ca, Sr, Ba) as nonlinear optical materials. Journal of the American Chemical Society. 2011;**133**(49):20001-20007

[78] Liang ML, Hu CL, Kong F, Mao JG. BiFSeO3: An excellent SHG material designed by aliovalent substitution. Journal of the American Chemical Society. 2016;**138**:9433-9436

[79] Kubelka P. Ein Beitrag zur Optik der Farbanstriche (Contribution to the optic of paint). Zeitschrift fur technische Physik. 1931;**12**:593-601

[80] Tauc J. Absorption edge and internal electric fields in amorphous semiconductors. Materials Research Bulletin. 1970;**5**(8):721-729

Chapter 6

Crystallization: Its Mechanisms and Pharmaceutical Applications

Hendrik J.R. Lemmer and Wilna Liebenberg

Abstract

The crystallization of small-molecule drugs plays an important role in the pharmaceutical industry. Since many downstream industrial processes are heavily influenced by the crystalline properties of a drug, that is, crystal shape, size distribution, and polymorphic form, control over the crystallization process can facilitate manufacturing and testing. However, before the crystallization process can be controlled, an understanding of its underlying mechanisms is required. In this chapter, we will look at the thermodynamic driving force behind crystallization and how crystal nucleation and growth rates can be used to control the properties of the resulting crystals. Throughout the chapter, we give examples of how these control approaches can be applied in pharmaceutical research and industry to obtain crystals with desired properties. We then finish this chapter with a look at crystallization from the amorphous state, which differs from crystallization from solution and is a relevant topic in pharmaceutical sciences, since the preparation of an amorphous solid is a popular approach to enhancing the solubility of a drug.

Keywords: crystallization, pharmaceutical, mechanisms, models, nucleation, growth, amorphous

1. Introduction

Crystallization plays an important role in the manufacture and purification of small-molecule active pharmaceutical ingredients (APIs). It is estimated that between 70 and 80% of all small-molecule, APIs have at least one crystallization step in their manufacturing processes [1, 2]. To facilitate other downstream processes, such as filtration, drying, dissolution testing, and formulation, it is often desirable to be able to consistently produce crystals with specific properties, such as crystal size distribution, crystal shape (habit), and polymorphic form. Such control over the crystallization process requires accurate descriptions of crystal nucleation and growth kinetics, as well as solubility, breakage, and agglomeration data [3]. Crystallization control has gained even more interest since the release of the United States Food and Drug Administration's (FDA) Process Analytical Technology (PAT) framework, which aims to improve efficiency in pharmaceutical development, manufacturing, and quality assurance through innovative process development, analysis, and control [4].

Designing and implementing controlled crystallization processes can be time-consuming, and usually involve trial-and-error, model-based, or model-free (direct-design) approaches [5]. Accurate kinetic descriptions of crystal nucleation and growth form the backbone of the model-based approach which, although more complex, has the potential to alleviate the labor-intensiveness and facilitate the optimization of the traditional trial-and-error approach [6]. The models used in the model-based approach can even be used together, to more accurately describe the underlying crystallization mechanism(s). For example, in a study by Quilló and coworkers [6], three different crystal growth models were fit to experimental data, to find the most likely crystallization mechanism. They found that the underlying crystallization mechanism could best be described by an additive combination of simultaneous Birth-and-Spread (B+S) and Burton-Cabrera-Frank (BCF) models, which will be discussed later in this chapter. An example of the model-free approach is direct nucleation control (DNC), which attempts to directly control the number of nuclei present in a system. Using DNC, Abu Bakar and coworkers [5] were able to produce glycine crystals from water–ethanol mixtures that were larger than those obtained from uncontrolled crystallization.

To implement the above-mentioned approaches, the investigator needs to be able to monitor the crystallization process. The instrumental technique used will depend on what aspect of the process one needs to monitor. For the model-based approach, one would typically be interested in monitoring the supersaturation decay during crystallization. This can be done using techniques, such as attenuated total reflectance-Fourier transform infrared (ATR-FTIR) or near infrared (NIR) [7]. If the goal is to monitor the evolution of crystal size distribution over time, for example in the DNC study mentioned above, a technique like focused beam reflectance measurement (FBRM) can be used [5, 7]. In most investigations, the above-mentioned instrumental techniques are used concurrently, to gain as much information about the crystallization process as possible.

The crystallization processes and monitoring techniques mentioned above are concerned with crystallization from solution, where a reduction in temperature and/or the addition of an antisolvent is used to create supersaturated conditions. Under supersaturated conditions, the solute molecules can come together to form small masses, or nuclei, through a process called nucleation. The addition of more solute molecules to these nuclei is called crystal growth. There is also another type of crystallization that is of particular importance in pharmaceutical sciences, and that is crystallization from the amorphous state.

In this chapter, we will discuss the theoretical background of crystal nucleation and growth, look at examples of how these theoretical models can be implemented practically and give examples of applications in pharmaceutical research. We will also look at crystallization from the amorphous state and discuss techniques that can be used to delay it.

2. The driving force behind crystallization

The thermodynamic driving force behind crystal nucleation and growth rates is a chemical potential difference (Eq. (1)):

$$\mu_i - \mu_i^{sat} = RT \ln \left(\frac{a_i}{a_i^{sat}} \right) \tag{1}$$

where μ_i and a_i are the chemical potential and thermodynamic activity of solute i, respectively, and the superscript "*sat*" represents the property at supersaturated conditions, R is the universal gas constant and T is the absolute temperature [6, 8]. Because the activity of a solute is hard to measure experimentally, it is usually substituted with the term $\gamma_i x_i$, where γ_i is the activity coefficient and x_i the mole fraction of solute i. Since mole fractions can be measured experimentally, Eq. (1) is often rearranged and expressed in terms of a supersaturation (σ) driving force [8], presented here in Eq. (2):

$$\ln(\sigma) = \ln\left(\frac{\gamma_i x_i}{\gamma_i^{sat} x_i^{sat}}\right) = \ln\left(\frac{a_i}{a_i^{sat}}\right) = \frac{\mu_i - \mu_i^{sat}}{RT} \tag{2}$$

The value of x_i^{sat} can sometimes be determined experimentally from equilibrium solubility studies. However, if the solubility extrema cannot be determined experimentally, it can be estimated from a variety of drug solubility models [9], for example, the Apelblat model [10], or the van't Hoff–Jouyban–Acree (VH–JA) model [6].

Because the activity coefficients still need to be estimated, many studies make the simplifying assumption of setting the activity coefficients in Eq. (2) equal to 1, thereby expressing the supersaturation driving force only as a mole fraction $\left(x_i/x_i^{sat}\right)$, or as the concentration ratios $\left(C_i/C_i^{sat}\right)$, or $(S-1)$ where $S = \left(C_i/C_i^{sat}\right)$ [11–13]. Although these simplifications of the supersaturation driving force have been used successfully, they do not necessarily generalize to all solutes and solvents, and one might need to find a way to estimate the activity coefficient of the solute. Fortunately, there is a way to estimate the denominator$\left(\gamma_i^{sat} x_i^{sat}\right)$ in Eq. (2) in one go.

When dealing with phase equilibria, the traditional thermodynamic reference is a supercooled melt of the pure solute compound. Using this reference, the activity of a solute can be obtained from the enthalpy of fusion at the temperature of interest. In its most general form, this is done through Eq. (3) [14]:

$$\ln\left(a_i^{sat}\right) = \ln\left(\gamma_i^{sat} x_i^{sat}\right) = \frac{\Delta H_f(T_m)}{R}\left(\frac{1}{T_m} - \frac{1}{T}\right) - \frac{1}{RT}\int_{T_m}^{T} \Delta C_p dT + \frac{1}{R}\int_{T_m}^{T} \frac{\Delta C_p}{T} dT \tag{3}$$

where $\Delta H_f(T_m)$ is the enthalpy of fusion at the melting temperature, T_m, T is the temperature of interest and ΔC_p is the difference in heat capacity between the supercooled melt, $C_p(l)$ and the solute's solid-state $C_p(s)$, presented here in Eq. (4):

$$\Delta C_p = C_p(l) - C_p(s) \tag{4}$$

A practical problem with Eq. (3) is that ΔC_p has to be integrated down from the melting temperature to the temperature of interest, but far below the melting temperature the thermodynamic properties of the supercooled melt are not experimentally accessible [14]. Looking at the right-hand side of Eq. (3), ΔC_p is the only physical property that is difficult to obtain experimentally. This has led to several simplifying assumptions regarding Eq. (3), all concerned with how to handle ΔC_p.

A common assumption is to completely ignore ΔC_p [14], resulting in Eq. (5):

$$\ln\left(\gamma_i^{sat} x_i^{sat}\right) = \frac{\Delta H_f(T_m)}{R}\left(\frac{1}{T_m} - \frac{1}{T}\right) \tag{5}$$

However, when working with temperature ranges normally used in pharmaceutical processes, some approximations obtained from Eq. (5) have shown to be inaccurate [15, 16].

Another common assumption is that ΔC_p is constant and can be approximated by the entropy of fusion at the melting temperature $\Delta S_f(T_m)$, yielding Eq. (6) [14, 17]:

$$\ln\left(\gamma_i^{sat} x_i^{sat}\right) = \frac{\Delta H_f(T_m)}{RT_m} \ln\left(\frac{T}{T_m}\right) \tag{6}$$

If we can experimentally determine the isobaric heat capacity of the solid, $C_p(s)$, at different temperatures below the melting temperature, as well as the isobaric heat capacity of the melt, $C_p(l)$, the assumption can be made that ΔC_p is constant and equal to its value at the melting temperature, $\Delta C_p(T_m)$, as illustrated in **Figure 1** [14]. Provided that the compound in question does not decompose upon melting, this assumption gives Eq. (7):

$$\ln\left(\gamma_i^{sat} x_i^{sat}\right) = \frac{\Delta H_f(T_m)}{R}\left(\frac{1}{T_m} - \frac{1}{T}\right) - \frac{\Delta C_p(T_m)}{R}\left(\ln\left(\frac{T_m}{T}\right) - \frac{T_m}{T} + 1\right) \tag{7}$$

If we have enough isobaric heat capacity data above the melting temperature, we can extrapolate down from $C_p(l)$ to the temperature of interest, see **Figure 1**, and use Eq. (3) in its general form [14]. In this case, we can use the linear dependency of ΔC_p on temperature to rewrite Eq. (4) as:

$$\Delta C_p = C_p(l) - C_p(s) = q + r(T_m - T) \tag{8}$$

where q and r are regression parameters, obtained from extrapolating down from $C_p(l)$ to temperatures of interest, calculating the difference in heat capacity between the extrapolated data and $C_p(s)$, and plotting these differences in heat capacity, $\Delta C_p(T)$, against $(T_m - T)$. Notice from the above that the value of q corresponds to $\Delta C_p(T_m)$ in Eq. (7). Once we have estimates of q and r, we can plug them into Eq. (3) and solve the integrals to give the most comprehensive estimate of $\left(\gamma_i^{sat} x_i^{sat}\right)$ in the form of Eq. (9):

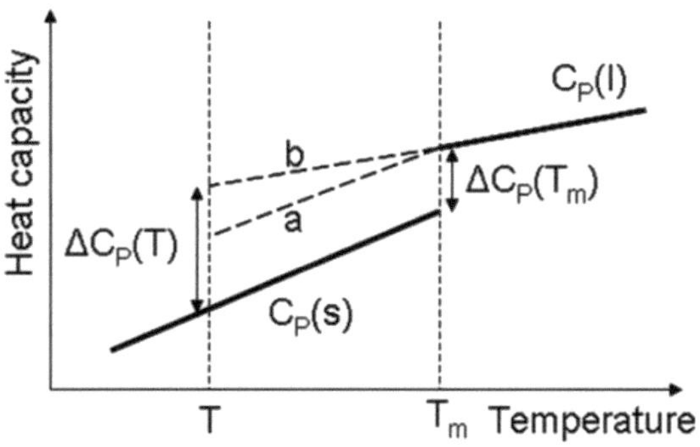

Figure 1.
Illustration of the experimental heat capacities (solid lines) and extrapolated heat capacities (dashed lines) used to approximate the heat capacity term in Eq. (3). Reproduced from ref. [14] with permission from Elsevier.

$$\ln\left(\gamma_i^{sat} x_i^{sat}\right) = \frac{\Delta H_f(T_m)}{R}\left(\frac{1}{T_m} - \frac{1}{T}\right) - \frac{q}{R}\left(\ln\left(\frac{T_m}{T}\right) - \frac{T_m}{T} + 1\right) - \frac{r}{R}\left(T_m \ln\left(\frac{T_m}{T}\right) - \frac{T_m^2}{2T} + \frac{T}{2}\right) \tag{9}$$

Once the values of $\left(\gamma_i^{sat} x_i^{sat}\right)$ at different temperatures have been properly estimated, the values of γ_i can be calculated from experimental solubility data.

3. Nucleation

In the introduction, we mentioned that crystallization consists of two processes, namely nucleation and growth. Depending on the rates of these two processes, the molecules that make up the crystal may pack differently, giving rise to different crystal structures, or polymorphic forms, of the same compound.

Nucleation is the first step in the crystallization process, and it consists of two mechanisms, namely homogeneous and heterogeneous nucleation. Homogeneous nucleation is triggered by spontaneous fluctuations in the density of the liquid, while heterogeneous nucleation is triggered by contact of the liquid with a foreign solid surface, like an impurity or metastable polymorph [7]. Since most industrial crystallization processes involve heterogeneous nucleation, we will focus mainly on this mechanism, which can be expressed mathematically as Eq. (10):

$$J = N_0 v \; exp\left(\frac{-\Delta G^* \Phi}{kT}\right) \tag{10}$$

where J is the number of nuclei formed per unit time per unit volume, N_0 is the number of solute molecules per unit volume, v is the frequency of molecular transport at the nucleus-liquid interface, Φ is the heterogeneous nucleation factor, a function of the contact angle between the nuclei and a foreign surface in the solution with values ranging from 0 to 1, and ΔG^* is the free energy barrier to nucleation of a sphere [7, 18], defined by:

$$\Delta G^* = \frac{16\pi v^2 \gamma_{sl}{}^3}{3(kT)^2(\ln\sigma)^2} \tag{11}$$

where v is the molecular volume of the solute, γ_{sl} is the solid–liquid interfacial tension per unit area, k is the Boltzmann constant and T and σ have the same meanings as before [18]. The equation for the rate of homogenous nucleation is similar to Eq. (10) but does not contain the heterogeneous nucleation factor, Φ, and has a different pre-exponential factor [18].

Looking at Eq. (10), we see some interesting mechanistic features of the nucleation step. First off, we can expect a higher number of nuclei to form from solvents in which the solute is more soluble because these will give higher values of N_0, which is a concentration term. Concretely, since the pre-exponential term, $N_0 v$ is an estimate of the probability of intermolecular collisions, and the term v is mainly determined by the degree of agitation which can be controlled to be constant between different experiments [18], higher solubility will lead to higher values of N_0, which will increase the likelihood of nuclei forming. The exponential term itself is a negative exponential, it will decay and asymptotically approach zero for large values in the exponential. In other words, we can expect fewer nuclei to form per unit volume of solute if the free energy barrier to nucleation, ΔG^*, is large. The free energy barrier to nucleation is itself dependent on the interfacial tension, γ_{sl}, expressed as:

$$\gamma_{sl} = 0.414kT(C_s N_A)^{2/3}\left(\ln\left(\frac{C_s}{C_{eq}}\right)\right) \tag{12}$$

where C_s is the ratio of the density of the solute to its molar mass, N_A is Avogadro's number, C_{eq} is the equilibrium solubility of the solute and k and T have the same meanings as above [19]. From Eq. (12), we see that γ_{sl} is inversely proportional to the logarithm of the equilibrium solubility, suggesting that higher values of C_{eq} will result in lower values of γ_{sl}. Putting it all together, we see that under conditions of higher solubility, C_{eq}, the interfacial tension, γ_{sl}, will be lower, lowering the energy barrier to nucleation, ΔG^*, and thereby increasing the number of nuclei formed, J.

In the introduction, we mentioned that there are model-free approaches to controlling crystallization and that DNC is an example of such an approach. In short, the DNC approach attempts to control the number of nuclei present by adding solvent or antisolvent, increasing or decreasing the temperature, or a combination of both, to manipulate the solute's solubility. Abu Bakar and coworkers employed DNC and were able to produce progressively larger glycine crystals if they lowered the number of newly formed particles [5].

Another example of where the nucleation step of crystallization was used as an intervention point to control the crystal size distribution, is in the work of Fujiwara and coworkers [20]. Paracetamol (acetaminophen), like many other small-molecule APIs, tends to form agglomerates during crystallization, especially if the crystals are small ($< 100\ \mu m$). This can lead to formulation-related problems further downstream. Fujiwara and coworkers used a solubility curve and determined the metastable limit, that is, the degree of supersaturation that corresponds to

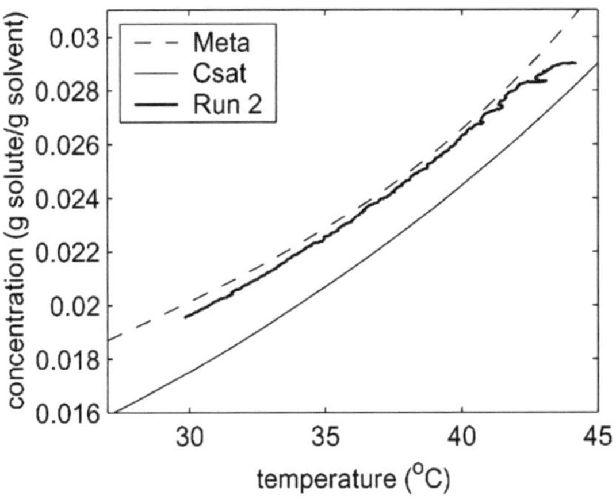

Figure 2.
Illustration of how the degree of supersaturation (Run 2) was controlled to fall between the metastable limit (Meta) and the solubility curve (Csat) during cooling. Reproduced from ref. [20] with permission from the American Chemical Society.

spontaneous nucleation. They found that if the degree of supersaturation in a seeded batch crystallization process can be temperature controlled during cooling to stay above saturation but below the metastable limit, as illustrated in **Figure 2**, larger paracetamol crystals with negligible nucleation and agglomeration can be obtained [20].

4. Crystal growth mechanisms

Once the nucleation step of crystallization is underway, more solute molecules can become incorporated into these nuclei and crystal growth can begin. Generally, crystal growth is considered to consist of two steps in series, namely volume diffusion and surface integration. Volume diffusion consists of the diffusion of solute molecules from the solution, through the boundary layer surrounding the crystal, to its surface. Surface integration can be split into three subprocesses, starting with the desolvation of a solute molecule unto the surface of the crystal, followed by the transfer of this molecule from its point of arrival across the crystal's surface. The latter is referred to as surface diffusion. Finally, the desolvated molecule reaches an energetically favorable location (kink site) on the crystal's surface and becomes incorporated into the crystal lattice [13]. The steps in crystal growth described above are illustrated in **Figure 3**.

In this section, we will discuss the different crystal growth mechanisms, their theoretical basis, and examples of how they are used in pharmaceutical sciences. The topic of surface energy models is quite extensive and will not be covered in detail in this section. However, it plays an important role in how different crystal growth rates can lead to different crystal habits and is, therefore, a relevant topic in solid-state pharmaceutical investigations, as it influences downstream processes, such as dissolution, powder flow, milling, granulation, and compaction. Therefore, we will start this section with a discussion on the Gibbs–Curie–Wulff theorem and then look at a popular surface energy model, namely the Bravais–Friedel–Donnay–Harker (BFDH) model.

Also, in this section, we will not discuss adsorption layer or diffusion-reaction models but will instead focus on the more popular power law, birth-and-spread, and screw dislocation models, and show how information regarding the rate-limiting steps in crystal growth can be inferred from the power-law model.

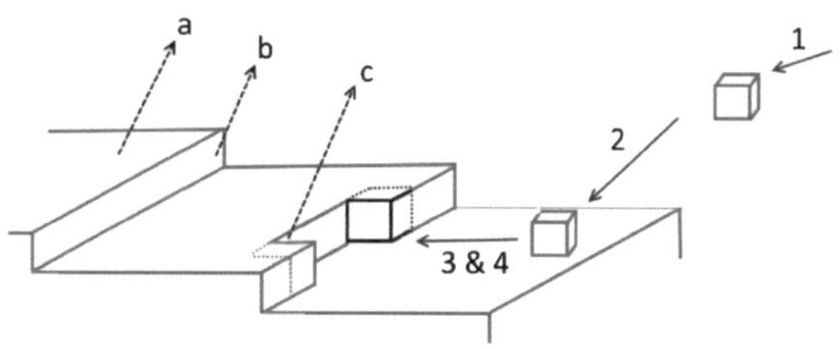

Figure 3.
A schematic representation of crystal growth. The steps are (1) the diffusion of a molecule from the solution to the crystal surface, (2) desolvation of the solute molecule unto the crystal surface, (3) transfer across the surface, and (4) incorporation of the molecule into the lattice. The numbering represents—(a) the terrace, (b) the step, and (c) the kink site. Reproduced from ref. [21] with permission from Springer Nature.

4.1 Surface energy models

The Gibbs–Curie–Wulff theorem originated when Gibbs postulated that, at equilibrium, a crystal should take on a form such that the product of its surface area and the surface-free energy is minimized. Later, Curie proposed that there exists a direct proportionality between the normal growth rates of crystal faces and surface-free energy. Wulff later stated that, at equilibrium, there exists a central point (the Wulff point) within a crystal such that the distances of the crystal faces from this point are proportional to the specific surface-free energies of those faces [22].

Li and coworkers [22] investigated whether organic crystals, with their diverse molecular shapes, noncovalent interactions, and hydrogen bonding, would still follow the Gibbs–Curie–Wulff theorem. Using a heteroacene molecule as crystal former, they found that not only did the shape and shape evolution of their equilibrium crystals follow the Gibbs–Curie–Wulff theorem, but that the packing of the molecules inside the crystals also reflected the principle of minimizing surface-free energy.

A crystal lattice can be thought of as a repeating 3D pattern consisting of a set of points such that each point has identical surroundings. A lattice has three spatial dimensions (a, b, c), also referred to as the lattice constants. In crystallography, Miller indices are used to express a crystallographic plane in terms of a 3D lattice using the notation (hkl). The physical meaning of Miller indices (hkl) is a set of parallel crystal facets with equivalent interplanar spacing [23]. The interplanar spacing is defined as the distance between two adjacent parallel planes (facets) with the same Miller indices, denoted as d_{hkl}. So, for a cubic lattice system, where the lattice constant is simply a, the interplanar spacing is defined as:

$$d_{hkl} = \frac{a}{\sqrt{h^2 + k^2 + l^2}} \tag{13}$$

And for a more complex tetragonal system [23, 24]:

$$d_{hkl} = \frac{1}{\sqrt{\left(\frac{h}{a}\right)^2 + \left(\frac{k}{b}\right)^2 + \left(\frac{l}{c}\right)^2}} \tag{14}$$

The BFDH model attempts to predict the crystal habit based on the growth rates of the crystal facets. Concretely, the BFDH model states that the growth rate of a facet is inversely proportional to that facet's interplanar spacing [21, 25]. Unlike the other crystal growth models that will be discussed later in this chapter, the BFDH model does not have a mathematical representation. Instead, predictions of crystal morphology based on the BFDH model are obtained from computer modeling software, such as Mercury 2020.1, Cerius^2, or Materials Studio.

Since the BFDH model does not explicitly take into account factors, such as molecular interactions but is instead concerned with the geometric aspects of the crystal, there have been instances where predictions made from the BFDH model were not in agreement with experimental observations. For example, while using the BFDH model to predict the morphologies of two carvedilol polymorphs, Prado and coworkers found that their BFDH model predicted thicker crystals of form II than were experimentally obtained and the model also predicted more faces for form III [26]. Similarly, Nichols and Frampton found that the BFDH model predicted morphologies for paracetamol polymorphic forms I and II that differed significantly from those observed experimentally [27]. Instead of using the BFDH model to predict morphologies, Turner and coworkers [28] used it to identify the important morphological faces of lovastatin crystals obtained from different solvents. They found that the calculated attachment energies of these important faces correlated reasonably well with the observed crystal morphologies.

4.2 Power-law model

The first mathematical model of crystal growth we will discuss is the power-law, or empirical, model. As the name suggests it is an empirical model, raised to some power, and as such it does not offer much insight into the actual growth mechanism. However, it is widely regarded for its flexibility and for being less complex and easier to use than the birth-and-spread or screw dislocation models.

Concretely, the power-law model expresses the growth rate (G_{PWR}) in terms of the degree of supersaturation as:

$$G_{PWR} = k_g(\ln(\sigma))^g \tag{15}$$

where the kinetics constant, k_g, is expected to follow Arrhenius temperature dependence, and is defined as:

$$k_g = k_{g0}\, exp\left(\frac{-E_a}{RT}\right) \tag{16}$$

and where g is the temperature-independent growth order parameter, k_{g0} is a pre-exponential factor, E_a is the activation energy and σ, R, and T have the same meanings as before [6].

We mentioned that the power-law model does not explicitly offer insights into the specific crystal growth mechanism. However, recently certain consistencies between the values of g and E_a and growth mechanisms have come to the fore. For instance, there is now evidence that for values of E_a ranging from 10–20 kJ/mol the rate-limiting step for crystal growth is volume diffusion, and for values ranging from 40–100 kJ/mol, surface integration [13, 17]. For values of g higher than unity, the rate-limiting step is also considered to be surface integration [13].

We also mentioned the flexibility of the power-law model. In a paper concerning the crystal growth kinetics of piracetam polymorphs, Soto and Rasmuson [13] substituted the supersaturation driving force in Eq. (15) with a different driving force, namely a mass balance expression based on Haüy's law, which states that one can assume a crystal's shape to be constant if nucleation, agglomeration, breakage and growth rate dispersion can all be assumed to be negligible. The driving force can then be expressed as Eq. (17):

$$\ln\sigma = \left(\Delta C_0 - \left[\left(\frac{\overline{L}}{\overline{L}_0}\right)^3 - 1\right]\frac{W_0}{M}\right)\Big/ C_i^{sat} \tag{17}$$

where $\Delta C_0 = C_i - C_i^{sat}$, $\overline{L}$ is the crystal length at any instant, $\overline{L}_0$ is the mean initial size of the seed crystals, W_0 is the mass of the seed crystals, M is the mass of solvent and C_i and C_i^{sat} have the same meanings as before [13]. Substituting back into Eq. (15) and combining with Eq. (16), they modeled the crystal growth rates using Eq. (18):

$$G_{PWR} = \frac{dL}{dt} = k_{g0}\,exp\left(\frac{-E_a}{RT}\right)\left[\left(\Delta C_0 - \left[\left(\frac{\overline{L}}{\overline{L}_0}\right)^3 - 1\right]\frac{W_0}{M}\right)\Big/ C_i^{sat}\right]^g \tag{18}$$

where dL/dt is the rate of change of a crystal's characteristic linear dimension with time and all the other terms have the same meanings as before [13]. Using Eq. (18), they were able to accurately model the growth rates of their piracetam polymorphs and from the values of g and E_a, they were able to determine that surface integration was the rate-limiting step in their crystals' growth.

4.3 Birth-and-spread model

The birth-and-spread model (B+S) models surface, or two-dimensional, nucleation which is believed to occur at intermediate levels of supersaturation. Under such conditions, surface nuclei can form at the edges, corners, and even faces of crystals, creating new growth steps [13]. These surface nuclei then spread to create a new crystalline layer. The B+S model is presented here in Eq. (19)–(22):

$$G_{B+S} = \tau_1(\sigma - 1)^{2/3}(\ln(\sigma))^{1/6}\,exp\left(\frac{-\tau_2}{T^2\ln(\sigma)}\right) \tag{19}$$

$$\tau_1 = \left(\frac{16}{\pi}\right)^{1/3} h^{1/6} D_{surf}\left(\beta'\frac{\Gamma^*}{x_s}\right)^{2/3}(V_m\Gamma N_A)^{5/6} \tag{20}$$

$$\tau_2 = \frac{\pi}{3} V_m h\left(\frac{\gamma_{sl}}{k}\right)^2 \tag{21}$$

$$D_{surf} = A_{surf}\,exp\left(\frac{-E_{a,surf}}{RT}\right) \tag{22}$$

where τ_1 and τ_2 are lumped model parameters, h is the step height, β' is a correction factor ($\beta' \leq 1$) , Γ is the molecular adsorption coverage of the solute and Γ^* is that property at equilibrium, x_s is the mean displacement of the adsorbed units over the surface, V_m is the molecular volume of the solute, D_{surf} is the surface diffusion coefficient, A_{surf} is a pre-exponential factor, $E_{a,surf}$ is the activation energy for surface diffusion and the rest of the parameters have the same meanings as before [6].

Despite being a very comprehensive model, there is a specific inconsistency between predicted data and experimental observation that should be pointed out, and that is that once growth from a surface nucleation site has spread all the way to the edge of the crystal face, forming a new crystalline layer, any further growth would require adsorption to a smooth surface. Since this kind of adsorption is energetically less favored than binding to dislocations, it would require high levels of supersaturation to overcome the energy barrier. However, experimental data have shown that crystal growth can occur at much lower degrees of supersaturation [8]. To address this inconsistency, a screw dislocation model can be used, which assumes that screw dislocations on a crystal surface serve as self-perpetuating growth sites.

4.4 Screw dislocation model

The Burton–Cabrera–Frank (BCF) model [29] is a screw dislocation model, that assumes that crystal growth stems from screw dislocations on a crystal surface and that growth from these ledges leads to more ledges in a self-perpetuating fashion, thereby attempting to explain how crystal growth can be obtained from supersaturations lower than that needed for the mechanism described by the B+S model. The BCF model can be expressed mathematically as Eqs. (23)–(25):

$$G_{BCF} = \frac{\tau_3 T}{\tau_4} (\ln(\sigma))^2 \tanh\left(\frac{\tau_4}{T \ln(\sigma)}\right) \tag{23}$$

$$\tau_3 = \frac{\Gamma^* D_{surf} V_m}{{x_s}^2} \tag{24}$$

$$\tau_4 = \frac{19 V_m \gamma_{sl}}{2k x_s} \tag{25}$$

where τ_3 and τ_4 are lumped parameters and all the other parameters have the same meanings as before [6].

Now that we have presented the B+S and BCF models, we can appreciate the simplicity of the power-law model. In the introduction, we mentioned that Quilló and coworkers [6] used the power law, B+S, and BCF models, separately and in different combinations, to model the crystal growth kinetics of a proprietary API and thereby attempted to ascertain the underlying growth mechanism(s). During the combination tests, they also tested for dominance of one model over the other, sudden mechanism crossover, and simultaneous mechanisms. They found that the growth process could be best described if both the B+S and BCF models were simultaneously active, in an additive manner, with a smooth crossover from initial surface nucleation dominated growth to screw dislocation (spiral growth) dominated growth.

5. Crystallization from the amorphous state

One of the ways to improve the solubility of a drug is to prepare an amorphous solid from it. The molecules inside these amorphous solids lack the long-range ordered packing of their crystalline counterparts, giving the amorphous solids higher thermodynamic activity and generally enhanced solubility. Methods to enhance the solubility of drugs have become a hot topic over recent years, and will likely remain a topic of interest, especially considering that approximately 36% of the 698 drugs currently available as immediate-release oral preparations and 60% of the 28912 new APIs under development are classified as biopharmaceutics classification system (BCS) class II or IV drugs [30, 31]. The BCS attempts to classify a drug into one of four categories, depending on its aqueous solubility and membrane permeability [32]. BCS class II and IV drugs have poor aqueous solubility but good membrane permeability, and poor aqueous solubility and poor membrane permeability, respectively. In other words, the market share of poorly soluble drugs is likely to increase in the future.

Currently, there are a limited number of amorphous preparations on the market [33]. A specific cause for concern is that the high thermodynamic activity of an amorphous solid, responsible for its enhanced solubility, also makes it thermodynamically unstable and likely to convert back to a more stable, but less soluble, crystalline form.
A mechanistic understanding of this conversion and the resulting glass-to-crystal (GC) growth is an important topic in pharmaceutical sciences.

GC growth rates cannot be readily explained by the thermodynamic driving force responsible for crystallization from solution, as described previously [34]. Initially, it was believed that crystallization from the amorphous state was dependent on the storage conditions, such as the storage temperature relative to the glass transition temperature (T_g), and the bulk properties of the amorphous solid, like α-relaxation processes [34–37]. However, it is now generally accepted that GC growth can take place at temperatures well below T_g and follows two mechanisms, namely fast surface crystal growth and slower growth in the bulk (interior) [31].

Both the surface and bulk crystallization processes are believed to be controlled by the self-diffusion of the molecules in the amorphous solid. Using indomethacin as model API, both Wu and Yu [38] and Swallen and Ediger [39] found that the crystal growth rates in the interior of their amorphous solids were proportional to the self-diffusion coefficients for temperatures close to T_g. It is believed that surface crystallization is so much faster than bulk crystallization, because of the increased molecular mobility at the surface. This was corroborated by Zhu and coworkers [40] who measured the surface smoothing of indomethacin glasses and found the self-diffusion on the surface to be at least one million times faster than in the interior.

With the mechanism of surface crystallization understood, immobilization of the molecules at the surface makes for an appealing target to delay the crystallization process. Yu and coworkers [41] coated indomethacin glasses with gold (10 nm) and two polyelectrolytes. They found that even a single layer of polyelectrolyte was enough to inhibit the growth of exiting crystals and that the molecular mobility of molecules at the surface of an amorphous solid can be sufficiently suppressed by a coating only a few nanometers thick. Their results suggest that nanocoating is a promising technique to stabilize amorphous solids.

6. Conclusions

The crystallization of small-molecule APIs plays an important role in the pharmaceutical industry. Because the properties of these crystals, that is, crystal habit, size distribution, and polymorphic form, influence many downstream processes, ranging from quality control testing to formulation; it is desirable to be able to consistently produce crystals with specific properties. To achieve this kind of control over the crystallization process, a thorough understanding of the underlying mechanisms is required. In this chapter, we looked at the thermodynamic driving force behind crystallization, and how it can be rearranged into a supersaturation driving force. Methods to estimate the degree of supersaturation were discussed. The mechanisms behind crystal nucleation and growth, and the mathematical models describing these mechanisms, were discussed and examples of how these mechanisms can be used as intervention points to control the properties of the resulting crystals were given. We ended this chapter with a look at amorphous solids, which have a natural tendency to crystallize back to a more stable, but less soluble form, and saw that immobilization of the molecules at the surface of these solids with even a single layer of the polymer was enough to stabilize the amorphous solid. With an increasing movement in the pharmaceutical industry toward streamlining manufacturing processes through control, the techniques discussed in this chapter might see even more general use in the future.

DOI: http://dx.doi.org/10.5772/intechopen.105056

Author details

Hendrik J.R. Lemmer* and Wilna Liebenberg
Centre of Excellence for Pharmaceutical Sciences, North-West University, Potchefstroom, South Africa

*Address all correspondence to: righard.lemmer@nwu.ac.za

References

[1] Lu J, Rohani S. Polymorphism and crystallization of active pharmaceutical ingredients (APIs). Current Medicinal Chemistry. 2009;**16**(7): 884-905. DOI: 10.2174/092986709787549299

[2] Reutzel-Edens SM. Achieving polymorph selectivity in the crystallization of pharmaceutical solids: basic considerations and recent advances. Current Opinion in Drug Discovery & Development. 2006; **9**:806-815

[3] Fytopoulos AA, Kavousanakis ME, Van Gerven T, Boudouvis AG, Stefanidis GD, Xiouras C. Crystal growth, dissolution, and agglomeration kinetics of sodium chlorate. Industrial & Engineering Chemistry Research. 2021; **60**(19):7367-7384. DOI: 10.1021/acs.iecr.1c00595

[4] FDA. Guidance for Industry PAT—A Framework for Innovative Pharmaceutical Development, Manufacturing, and Quality Assurance. 2022. Available from: https://www.fda.gov/media/71012/download [Accessed: March 22, 2022]

[5] Abu Bakar MR, Nagy ZK, Saleemi AN, Rielly CD. The impact of direct nucleation control on crystal size distribution in pharmaceutical crystallization processes. Crystal Growth and Design. 2009;**9**(3): 1378-1384

[6] Quilló GL, Bhonsale S, Gielen B, Van Impe JF, Collas A, Xiouras C. Crystal growth kinetics of an industrial active pharmaceutical ingredient: Implications of different representations of supersaturation and simultaneous growth mechanisms. Crystal Growth & Design. 2021;**21**(9):5403-5420

[7] Yu LX, Lionberger RA, Raw AS, D'Costa R, Wu H, Hussain AS. Applications of process analytical technology to crystallization processes. Advanced Drug Delivery Reviews. 2004; **56**(3):349-369. DOI: 10.1016/j.addr.2003.10.012

[8] Nagy ZK, Fujiwara M, Braatz RD, Myerson AS, Erdemir D, Lee AY. Monitoring and advanced control of crystallization processes. Handbook of Industrial Crystallization. 2019;**3**. DOI: 10.1017/9781139026949.011

[9] Jouyban A. Review of the cosolvency models for predicting solubility of drugs in water-cosolvent mixtures. Journal of Pharmacy & Pharmaceutical Sciences. 2008;**11**(1):32-58. DOI: 10.18433/J3PP4K

[10] Schall JM, Capellades G, Myerson AS. Methods for estimating supersaturation in antisolvent crystallization systems. CrystEngComm. 2019;**21**(38):5811-5817. DOI: 10.1039/C9CE00843H

[11] Tilbury CJ, Doherty MF. Modeling layered crystal growth at increasing supersaturation by connecting growth regimes. AIChE Journal. 2017;**63**(4): 1338-1352. DOI: 10.1002/aic.15617

[12] Schall JM, Mandur JS, Braatz RD, Myerson AS. Nucleation and growth kinetics for combined cooling and antisolvent crystallization in a mixed-suspension, mixed-product removal system: Estimating solvent dependency. Crystal Growth & Design. 2018;**18**(3): 1560-1570. DOI: 10.1021/acs.cgd.7b01528

[13] Soto R, Rasmuson ÅC. Crystal growth kinetics of Piracetam polymorphs in ethanol and isopropanol.

Crystal Growth & Design. 2019;**19**(8): 4273-4286. DOI: 10.1021/acs.cgd.8b01733

[14] Nordström FL, Rasmuson ÅC. Determination of the activity of a molecular solute in saturated solution. The Journal of Chemical Thermodynamics. 2008;**40**(12): 1684-1692. DOI: 10.1016/j.jct.2008.06.016

[15] Neau SH, Flynn GL. Solid and liquid heat capacities of n-alkyl para-aminobenzoates near the melting point. Pharmaceutical Research. 1990;**7**(11): 1157-1162. DOI: 10.1023/A:1015984310068

[16] Hojjati H, Rohani S. Measurement and prediction of solubility of paracetamol in water– isopropanol solution. Part 2. Prediction. Organic Process Research & Development. 2006; **10**(6):1110-1118. DOI: 10.1021/op060074g

[17] Moffitt Schall J, Myerson AS. Solutions and Solution Properties. In: Myerson AS, Erdemir D, Lee AY, editors. Handbook of Industrial Crystallization. 3rd ed. Cambridge: Cambridge University Press; 2019. pp. 1-31. DOI: 10.1017/9781139026949.001

[18] Gu CH, Young V Jr, Grant DJ. Polymorph screening: Influence of solvents on the rate of solvent-mediated polymorphic transformation. Journal of Pharmaceutical Sciences. 2001;**90**(11): 1878-1890. DOI: 10.1002/jps.1137

[19] Mersmann A. Calculation of interfacial tensions. Journal of Crystal Growth. 1990;**102**(4):841-847. DOI: 10.1016/0022-0248(90)90850-K

[20] Fujiwara M, Chow PS, Ma DL, Braatz RD. Paracetamol crystallization using laser backscattering and ATR-FTIR spectroscopy: metastability, agglomeration, and control. Crystal Growth & Design. 2002;**2**(5):363-370. DOI: 10.1021/cg0200098

[21] Hadjittofis E, Isbell MA, Karde V, Varghese S, Ghoroi C, Heng JY. Influences of crystal anisotropy in pharmaceutical process development. Pharmaceutical Research. 2018;**35**(5): 1-22. DOI: 10.1007/s11095-018-2374-9

[22] Li R, Zhang X, Dong H, Li Q, Shuai Z, Hu W. Gibbs–Curie–Wulff theorem in organic materials: a case study on the relationship between surface energy and crystal growth. Advanced Materials. 2016;**28**(8): 1697-1702. DOI: 10.1002/adma.201504370

[23] Sun S, Zhang X, Cui J, Liang S. Identification of the Miller indices of a crystallographic plane: a tutorial and a comprehensive review on fundamental theory, universal methods based on different case studies and matters needing attention. Nanoscale. 2020; **12**(32):16657-16677. DOI: 10.1039/D0NR03637D

[24] Askeland DR, Fulay PP, Bhattacharya DK. Essentials of Materials Science and Engineering. FL, USA: Cengage Learning. Inc.; 2010. pp. 358-364

[25] Donnay JH, Harker D. A new law of crystal morphology extending the law of Bravais. American Mineralogist: Journal of Earth and Planetary Materials. 1937; **22**(5):446-467

[26] Prado LD, Patricio BF, Gonçalves KM, Santos AB, Bello ML, Rocha GM, et al. Pharmaceutical material engineering: Evaluation of carvedilol polymorphs II and III surface by packing, modeling, and atomic force

measurements. Crystal Growth & Design. 2020;**20**(12):7901-7909. DOI: 10.1021/acs.cgd.0c01172

[27] Nichols G, Frampton CS. Physicochemical characterization of the orthorhombic polymorph of paracetamol crystallized from solution. Journal of Pharmaceutical Sciences. 1998;**87**(6):684-693. DOI: 10.1021/js970483d

[28] Turner TD, Hatcher LE, Wilson CC, Roberts KJ. Habit modification of the active pharmaceutical ingredient lovastatin through a predictive solvent selection approach. Journal of Pharmaceutical Sciences. 2019;**108**(5): 1779-1787. DOI: 10.1016/j.xphs.2018.12.012

[29] Burton WK, Cabrera N, Frank FC. The growth of crystals and the equilibrium structure of their surfaces. Philosophical Transactions of the Royal Society of London. Series A, Mathematical and Physical Sciences. 1951;**243**(866):299-358. DOI: 10.1098/rsta.1951.0006

[30] Benet LZ. The role of BCS (biopharmaceutics classification system) and BDDCS (biopharmaceutics drug disposition classification system) in drug development. Journal of Pharmaceutical Sciences. 2013;**102**(1):34-42. DOI: 10.1002/jps.23359

[31] Wang K, Sun CC. Crystal growth of celecoxib from amorphous state: polymorphism, growth mechanism, and kinetics. Crystal Growth & Design. 2019; **19**(6):3592-3600. DOI: 10.1021/acs.cgd.9b00597

[32] Amidon GL, Lennernäs H, Shah VP, Crison JR. A theoretical basis for a biopharmaceutic drug classification: the correlation of in vitro drug product dissolution and in vivo bioavailability. Pharmaceutical Research. 1995;**12**(3): 413-420. DOI: 10.1023/A:1016212804288

[33] Zhang SW, Yu L, Huang J, Hussain MA, Derdour L, Qian F, et al. A method to evaluate the effect of contact with excipients on the surface crystallization of amorphous drugs. AAPS PharmSciTech. 2014;**15**(6): 1516-1526. DOI: 10.1208/s12249-014-0178-8

[34] Xi H, Sun Y, Yu L. Diffusion-controlled and diffusionless crystal growth in liquid o-terphenyl near its glass transition temperature. The Journal of Chemical Physics. 2009;**130**(9): 094508. DOI: 10.1063/1.3081047

[35] Hikima T, Adachi Y, Hanaya M, Oguni M. Determination of potentially homogeneous-nucleation-based crystallization in o-terphenyl and an interpretation of the nucleation-enhancement mechanism. Physical Review B. 1995;**52**(6):3900. DOI: 10.1103/PhysRevB.52.3900

[36] Hatase M, Hanaya M, Oguni M. Studies of homogeneous-nucleation-based crystal growth: significant role of phenyl ring in the structure formation. Journal of Non-Crystalline Solids. 2004; **333**(2):129-136. DOI: 10.1016/j.jnoncrysol.2003.10.010

[37] Sun Y, Xi H, Chen S, Ediger MD, Yu L. Crystallization near glass transition: transition from diffusion-controlled to diffusionless crystal growth studied with seven polymorphs. The Journal of Physical Chemistry B. 2008; **112**(18):5594-5601. DOI: 10.1021/jp7120577

[38] Wu T, Yu L. Origin of enhanced crystal growth kinetics near Tg probed with indomethacin polymorphs. The Journal of Physical Chemistry B. 2006;

110(32):15694-15699. DOI: 10.1021/jp062771g

[39] Swallen SF, Ediger MD. Self-diffusion of the amorphous pharmaceutical indomethacin near Tg. Soft Matter Journal. 2011;7(21): 10339-10344. DOI: 10.1039/C1SM06283B

[40] Zhu L, Brian CW, Swallen SF, Straus PT, Ediger MD, Yu L. Surface self-diffusion of an organic glass. Physical Review Letters. 2011;**106**(25): 256103. DOI: 10.1103/PhysRevLett.106.256103

[41] Wu T, Sun Y, Li N, de Villiers MM, Yu L. Inhibiting surface crystallization of amorphous indomethacin by nanocoating. Langmuir. 2007;**23**(9): 5148-5153. DOI: 10.1021/la070050i

Section 3

Chirality

Chapter 7

Chiral Ice Crystals in Space

Akira Kouchi, Takashi Shimonishi, Tomoya Yamazaki, Masashi Tsuge, Naoki Nakatani, Kenji Furuya, Hiromasa Niinomi, Yasuhiro Oba, Tetsuya Hama, Hiroyasu Katsuno, Naoki Watanabe and Yuki Kimura

Abstract

We observed the formation of CO, CH_3OH, and H_2O ices using a cryogenic transmission electron microscope, to determine if chiral ice crystals could form under the conditions of interstellar molecular clouds and young stellar objects (protoplanetary disks) and to clarify the crystalline structure of these ices. Our results suggest that the following ice crystals are chiral: crystalline CO (α-CO) formed on amorphous H_2O (a-H_2O) grains in a 10-K molecular cloud, crystalline CH_3OH formed by the heating of amorphous CH_3OH on a-H_2O grains at 40–60 K in young stellar objects, and several polymorphs of hydrogen-ordered cubic ice crystals formed by the heating of a-H_2O at 80–100 K and direct condensation at 120–140 K in protoplanetary disks. We also investigated candidates for other chiral ices using published data. We found that NH_3 I and $NH_3{\cdot}H_2O$ I are chiral at low temperature and pressure conditions. If one-handed circularly polarized light is irradiated during the nucleation of these chiral ice crystals, homochiral crystals can be formed. These results have important implications for the origin of interstellar organic molecule homochirality.

Keywords: ice crystals, chirality, CO, H_2O, CH_3OH, NH_3, interstellar molecular cloud, protoplanetary disk, circularly polarized light, asymmetric nucleation

1. Introduction

The origin of biomolecular homochirality is one of the most important mysteries of the origin of life. However, asymmetric adsorption and/or asymmetric synthesis on inorganic crystal surfaces is a possible candidate for chiral selection [1–3]. Quartz and cinnabar are regarded as chiral crystals, and it has been suggested that the surfaces of achiral crystals (e.g., gypsum, calcite, and alkali feldspar) can act as chiral faces [2, 4]. These minerals, as both chiral and achiral crystals, could be formed in evolved bodies, such as meteoritic parent bodies and terrestrial planets; however, it is implicitly considered that there were/are no chiral crystals in interstellar grains. Using transmission electron microscopy (TEM), we demonstrated that chiral crystalline CO (α-CO) would form on icy grains in interstellar molecular clouds [5]; therefore, α-CO in molecular clouds could be regarded as the first chiral crystal in space. To build on this finding, we searched for other chiral ices in space via further laboratory experiments and literature searches. We used the term "ice" to describe a solid at low temperatures (e.g., H_2O, CO_2, CO, NH_3, CH_3OH, and their hydrates).

In this chapter, after a brief explanation of icy grains in space in Section 2, we describe the crystal structures of each chiral ice in Section 3. In Section 4, a formation mechanism for homochiral ice crystals in space is discussed. In Section 5, we suggest further areas of study for the determination of the origin of the homochirality of organic molecules on icy grains in space.

2. Icy grains in space

2.1 The evolution of icy grains during the formation of the solar system

The evolution of icy grains, from an interstellar molecular cloud to a solar system, is schematically illustrated in **Figure 1**. In 10-K interstellar molecular cloud, icy grains were composed of an amorphous silicate (a-silicate) core, an inner organic mantle, an outer icy mantle of amorphous H_2O (a-H_2O), and α-CO attached to a-H_2O mantle [6]. The composition of ice differs among molecular clouds, as shown in **Table 1**. The molecular cloud collapsed by gravitational contraction to form protosolar nebulas, during which the icy grains were heated according to their heliocentric distance. In the inner region, the grains were completely sublimated. However, in the outer region, some grains survived. Subsequent cooling led to the formation of crystalline silicates in the inner region and H_2O ice crystals in the Jovian region. The aggregation of these

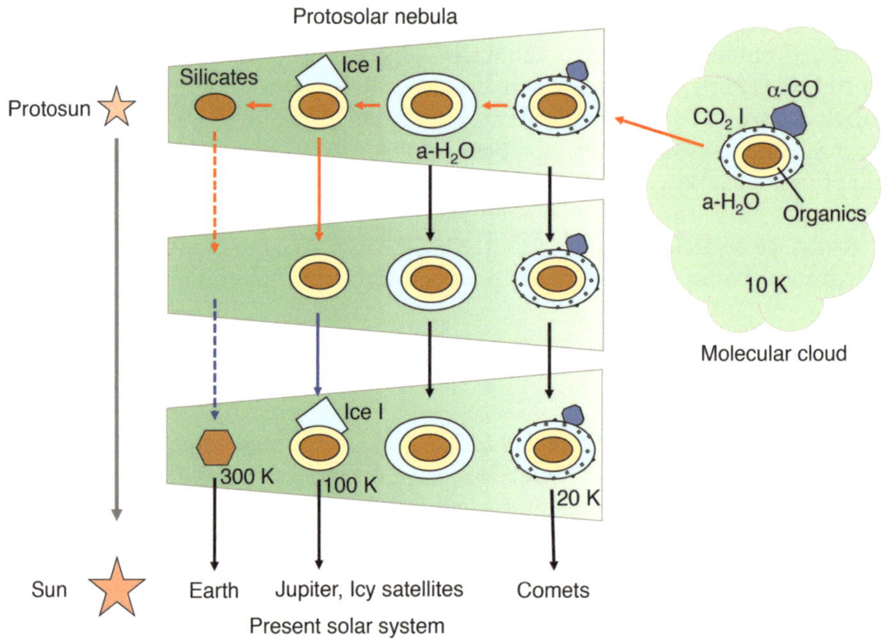

Figure 1.
Evolution of icy grains in space (grain sizes < 1μm). The compositions of the grains are indicated by different colors: Blue: CO, pale blue: H_2O, yellow: Organic matter, and brown: Silicate. Note that CH_3OH, NH_3, and minor components have been omitted for simplicity. Oval and polygonal grain forms are amorphous and crystalline, respectively. The density of the nebular gas, mainly composed of H_2, is indicated by the intensity of the green filling. The red and blue arrows indicate heating and cooling, respectively.

Molecule	MCs	MYSOs	LYSOs	Comets
H_2O	100	100	100	100
CO	9–67	3–26	(<3)–85	0.4–30
CO_2	14–43	11–27	12–50	4–30
CH_3OH	(<1)–12	(<3)–31	(<1)–25	0.2–7
NH_3	<7	~7	3–10	0.2–1.4
CH_4	<3	1–3	1–11	0.4–1.6
H_2CO	n.d.	2–7	~6	0.11–1
HCOOH	<2	(<0.5)–6	(<0.5)–4	0.06–0.14
NH_4^+	4–13	9–34	4–25	n.d.

Table 1.
Composition of ice in molecular clouds [7], young stellar objects [7], and comets [8] relative to H_2O. Abbreviations: MC, molecular cloud; MYSO, massive young stellar object; LYSO, low-mass young stellar object; n.d., no data.

grains led to the formation of planets via planetesimals, and remnant planetesimals from this outer region are the comets we observe today.

2.2 Infrared observation of ices

Information about the composition and crystallinity of icy grains can be gained from infrared (IR) astronomical observations. **Table 1** lists the main components of icy grains observed in molecular clouds and young stellar objects [7], including comets [8]. The most abundant component for all the objects is H_2O. The next most abundant components are CO and CO_2, although the abundance of CO varies depending on the object. For all the objects, the abundance of CH_3OH relative to H_2O ranges from lower than the detection limit to ~30%. Because CH_3OH can be formed from CO via the H-atom addition reaction on icy grains [9], it is suggested that the amount of CH_3OH reflects the evolutionary stage of objects. Although NH_3 is not detected in molecular clouds, it is detected in young stellar objects, while considerable amounts of NH_4^+ are tentatively assigned to all the objects [7]. It should be noted that the composition of cometary ices is quantitatively consistent with that of interstellar ices, suggesting an interstellar origin for cometary ices [8]. Among the crystals of these abundant molecules, possible chiral crystal candidates are H_2O, CO, CH_3OH, NH_3, and their hydrates, which will be discussed in the following section.

The comparison of astronomically observed and laboratory-measured IR spectra provides us with information on the crystallinity of ices, both amorphous and crystalline. H_2O ice is easily identified because of the spectral feature of the OH stretching mode around 3 μm, which differs between amorphous and crystalline H_2O ices [10]. The observed features of a molecular cloud (Elias 16) and a circumstellar envelope of an evolved star (OH231.8 + 4.2) could be fitted by a-H_2O at 23 K and crystalline H_2O ice at 77 K, respectively [11]. For a young stellar object (Orion BN), the observed feature could be fitted by a mixture of a-H_2O at 23 and 77 K and crystalline H_2O ice at 150 K [11]. These results, consistent with a theoretical study [12], are reflected in the crystallinity of the H_2O ice depicted in **Figure 1**.

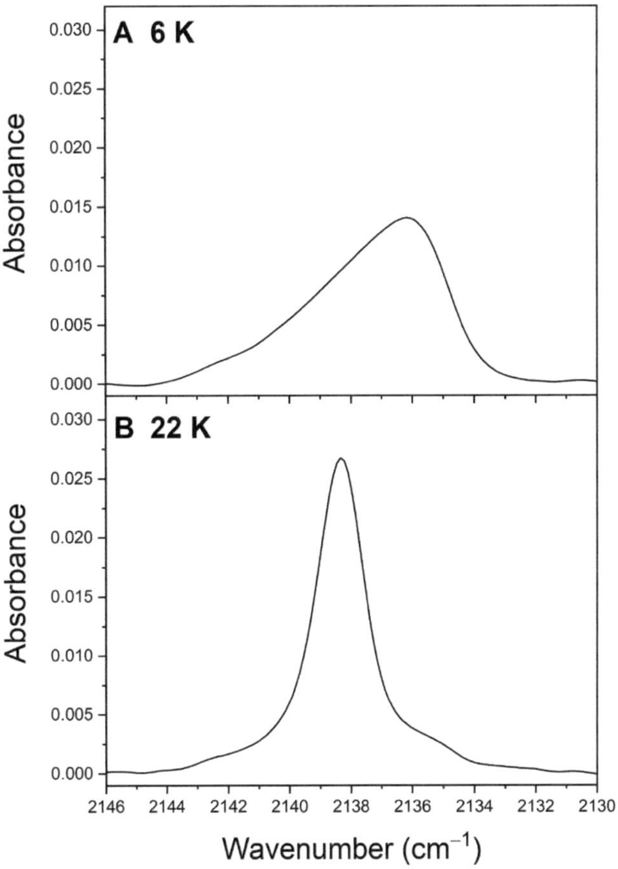

Figure 2.
Infrared spectra of solid CO. The CO gas was vapor deposited onto a Si(111) substrate at 6 K. The pressure of the chamber during deposition was 7.1×10^{-6} Pa (base pressure $< 2 \times 10^{-7}$ Pa), corresponding to the flux of 2×10^{13} molecules cm^{-2} s^{-1}, and the deposited amount after 60 minutes deposition was estimated to be 1.5×10^{17} molecules cm^{-2}. The deposited sample was warmed up stepwise with an increment of 2 K. The IR spectra were measured with a transmission configuration. The spectra measured at 6 K (a-CO) and 22 K (α-CO) are shown in (A) and (B), respectively.

Figure 2 shows the IR spectra of a-CO and α-CO measured by us. The sample deposition was done at a very low temperature (6 K) with a slow deposition rate (2×10^{13} molecules cm^{-2} s^{-1}), ensuring that the produced CO ice sample is amorphous [5]. The IR spectrum measured just after deposition shows an asymmetric feature with a peak near 2136 cm^{-1}: the IR spectrum of a-CO. During warming up to 22 K, the band shape gradually changed. The IR spectrum measured at 22 K (**Figure 2B**) shows a rather symmetric feature with a peak near 2138 cm^{-1}: the IR spectrum of α-CO. Recently, He et al. [13] reported the IR spectra of solid CO measured with a reflection-absorption IR spectrometry. They observed a very slight change in the peak position (~1 cm^{-1}) during warm-up and attributed this change to the phase transition from a-CO to α-CO. However, it should be noted that determination of crystallinity based on a reflection-absorption IR spectrometry measurement tends to be difficult and it is probable that their ice sample after deposition could be a mixture of a-CO and α-CO. Thus, we consider that the spectra shown in **Figure 2** are the first IR spectra of "pure" a-CO and α-CO measured in a laboratory. It is expected

Species	Phase	Space group	H-order	T (K)	Method [Ref]
H_2O	XIh[a]	$P2_1$	O		T [18]
	XIh[b]	$P2_12_12_1$	O		T [18]
	XIc[c]	$P2_1$	O		T [18]
	XIc[d]	$P2_12_12$	O		T [18]
	XIc[e]	$P4_12_12$	O		T [18]
	XIc[f]	$P4_1$	O		T [19]
	III[g]	$P4_12_12$	D	250	N [20]
CO	α	$P2_13$	n/a	14–30	N [21], E [5]
CH_3OH	α	$P2_12_12_1$	O	15, 110	N [22], E [23]
NH_3	I	$P2_13$	O	77	X [24], N [25]
$NH_3{\cdot}H_2O$	I	$P2_12_12_1$	O	~110	X [24], N [26]
$NH_3{\cdot}2H_2O$	I[h]	$P2_12_12_1$	O	100	X [17]
		$P2_13$	PO	4–174	N [26, 27]
CH_4	II[h]	$P2_13$	O	12, 25	I [28]
		$P2_13$	O		T [29]
		$Fm\bar{3}c$	PO	24.5	N [30]

[a]*Hydrogen-ordered hexagonal ices called H12–H15 by Raza et al. [18].*
[b]*Hydrogen-ordered hexagonal ices called H6 and H7 by Raza et al. [18].*
[c]*Hydrogen-ordered cubic ices called C10 and C11 by Raza et al. [18].*
[d]*Hydrogen-ordered cubic ice called C7 by Raza et al. [18].*
[e]*Hydrogen-ordered cubic ice called C2 by Raza et al. [18].*
[f]*Hydrogen-ordered cubic ices called d by Geiger et al. [19].*
[g]*High-pressure ice, measured at 0.28 GPa.*
[h]*Two space groups have been proposed.*

Table 2.
Candidate chiral ice crystals in molecular clouds and protoplanetary disks. Abbreviations: O, order; D, disorder; PO, partial order; T, theory; N, neutron diffraction; X, X-ray diffraction; E, electron diffraction; I, far-infrared spectroscopy.

that a comparison of these laboratory spectra with astronomical observations will be made in the near future, which will further the discussion of the crystallinity of solid CO in molecular clouds.

The laboratory-measured spectra of amorphous and crystalline CH_3OH phases differ [14]; however, because the astronomically observed spectra of the OH stretching modes of CH_3OH overlap with those of H_2O, it is difficult to obtain information about the crystallinity of CH_3OH. Zanchet et al. [15] measured the near- and mid-IR spectra of amorphous and crystalline NH_3 at 15 and 85 K, respectively, and found that both spectra were similar, except for a band around 1100 cm^{-1} [15], which demonstrates the difficulty of obtaining information on the crystallinity of NH_3. At 83 K, the measured IR spectra of the amorphous and crystalline phases of $NH_3{\cdot}H_2O$ differ between 700 and 1100 cm^{-1} [16], and only a crystalline phase has been measured for $NH_3{\cdot}2H_2O$ at 100 K [17]. However, it is expected that a comparison of these laboratory spectra with astronomical observations will be made in the near future.

3. The crystal structures of ices at low temperatures and pressures

Table 2 lists chiral ice crystal candidates in molecular clouds and protoplanetary disks, and the crystal structures of the respective species are described based on our observations and published data.

3.1 CO

α-CO is a thermodynamically stable phase of solid CO at low temperatures [31]. We observed the morphology of CO deposited on a-H_2O by TEM, as shown in **Figure 3** [5]. The CO formed three-dimensional polyhedral crystals, and the diffraction pattern confirmed that the CO crystals were α-CO. It has long been debated whether the crystal structure of α-CO is an orientationally ordered phase (space group: $P2_13$) or a disordered one (space group: $Pa\bar{3}$). Wang et al. [21] confirmed by neutron diffraction that α-CO belongs to $P2_13$, as proposed by Vegard [32]. Therefore, we concluded that α-CO in molecular clouds would be chiral.

3.2 H_2O

3.2.1 Ice Ih, Ic, and XI

Hydrogen-disordered hexagonal ice (ice Ih) is a thermodynamically stable phase of solid H_2O under low pressures at temperatures >72 K [33]. At temperatures <72 K, hydrogen-ordered ice XI becomes the thermodynamically stable phase [34, 35]. It is widely accepted that doping (e.g., KOH) is essential for the formation of ice XI at low temperatures [36]. A thermodynamically metastable phase of hydrogen-disordered cubic ice (ice Ic) also exists at temperatures between 100 and 200 K [33]. The crystal structures of ice Ih and ice Ic are very similar except for the stacking sequences of their layers: ice Ih is ABABAB and ice Ic is ABCABC. The space groups of ice Ih, ice XI, and ice Ic are $P6_3/mmc$, $Cmc2_1$, and $Fd\bar{3}m$, respectively, and all these crystals are not chiral.

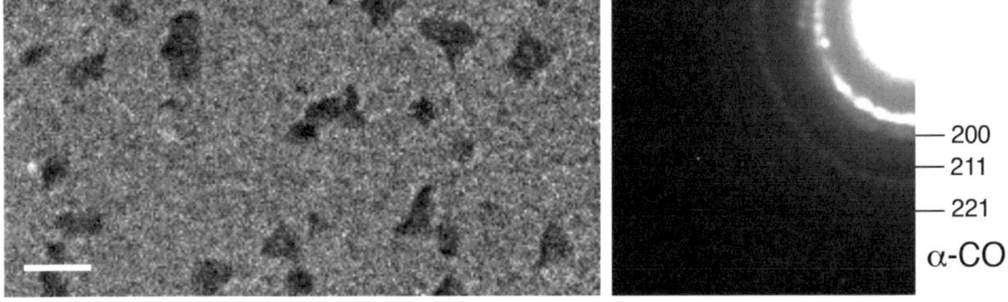

Figure 3.
Transmission electron microscopy image of (A) crystalline CO (α-CO) deposited on amorphous H_2O at 19 K, and (B) the electron diffraction pattern of α-CO on amorphous Si at 19 K. White scale bar = 500 nm.

3.2.2 Hydrogen-ordered cubic ice

Although the existence of hydrogen-ordered cubic ice (ice XIc) has been discussed theoretically [18, 19, 37–39], there has been no experimental evidence for this crystal structure. Raza et al. [18] and Geiger et al. [19] suggested 11 and 4 different structures, respectively, for ice XIc. We observed the annealing of ice Ic deposited on an a-SiN thin film by TEM and found that several polymorphs of ice were formed at temperatures between 100 and 130 K without doping [40]. **Figure 4** represents the TEM images and corresponding electron diffraction patterns of the ice XIc formed by the annealing of a-H_2O and ice Ic, showing the formation of ice XIc. However, we could not determine which structures were formed in terms of the different structures proposed by Raza et al. [18] and Geiger et al. [19]. However, because five-twelfths of the proposed structures were chiral (space group: $P4_12_12$, $P2_12_12$, $P2_1$, and $P4_1$), some chiral crystals might be included not only in our samples but also in H_2O ice crystals in

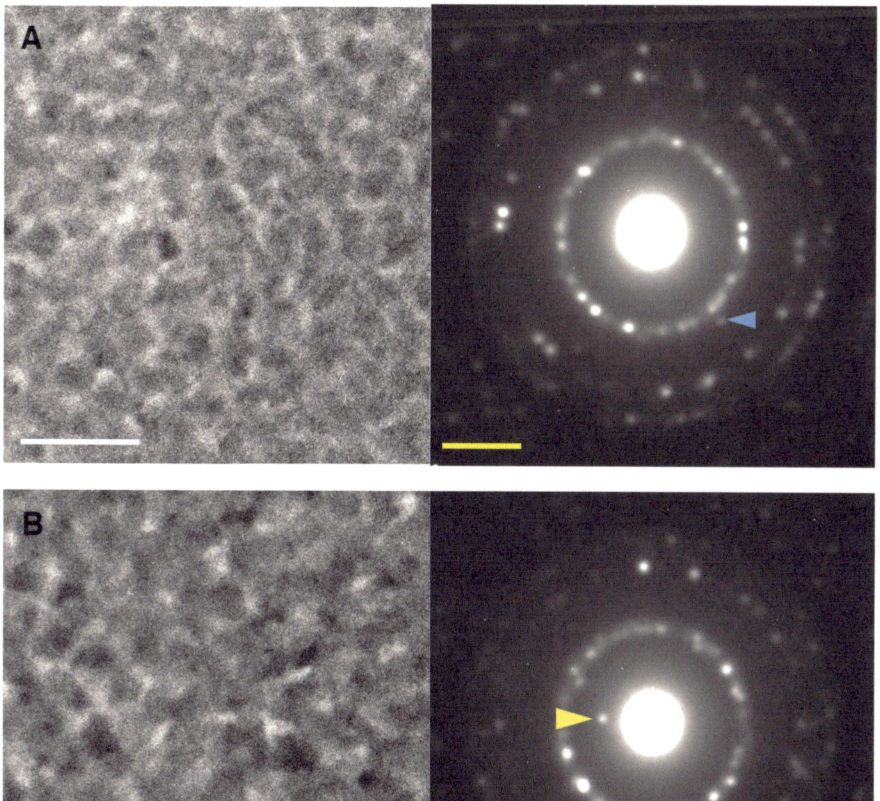

Figure 4.
Transmission electron microscopy images and corresponding electron diffraction patterns of hydrogen-ordered cubic ice (ice XIc) formed by (A) the annealing of amorphous H_2O at 130 K and (B) the annealing of ice Ic at 130 K. The blue and yellow arrowheads indicate the diffraction spots of d = 4.50 Å *and* d = 6.41 Å*, respectively, originating from ice XIc. The white and yellow scale bars are 500 nm and 2 nm^{-1}, respectively.*

space. Furthermore, the calculation of the infrared spectra of ice XIc, as demonstrated by Geiger et al. [19], will help identify ice XIc in space.

3.2.3 Hydrogen-ordered hexagonal ice

As mentioned in Section 3.2.1, the thermodynamically stable phase of hydrogen-ordered hexagonal ice is ice XI. However, Raza et al. [18] proposed 15 different structures of hydrogen-ordered hexagonal ice (XIh) as metastable phases, and seven-fifteenths of the proposed structures are chiral (space group: $P2_12_12_1$, $P2_1$, and $P1$). Although there has been no experimental investigation of this, we should consider the occurrence of these polymorphs, as demonstrated in the Section 3.2.2.

3.2.4 Ice III

Here, it is worthwhile commenting on the structure of ice III, although ice III is stable only at higher pressures between 210 and 344 MPa and higher temperatures between 238 and 256 K [33]. The space group of ice III is $P4_12_12$, meaning that the arrangement of the oxygen atoms is ordered and chiral. Conversely, the arrangement of the hydrogen atoms is disordered [20]. Therefore, we concluded that the surface of ice III does not behave as an asymmetric catalyst on which asymmetric adsorption and/or asymmetric synthesis can proceed.

3.3 CH_3OH

α-CH_3OH is a thermodynamically stable phase of solid CH_3OH at temperatures < 157 K under low pressure [41]. Torrie et al. [22] showed by neutron diffraction that α-CH_3OH is chiral, including the positions of the hydrogen atoms (space group: $P2_12_12_1$). Furuya et al. [23] observed the deposition of CH_3OH on a-H_2O using TEM between 90 and 120 K and found that α-CH_3OH was formed at temperatures >100 K; however, they did not observe the crystallization of amorphous CH_3OH (a-CH_3OH). Using TEM, we observed the formation of α-CH_3OH crystals during the warming of a-CH_3OH deposited on a-H_2O (**Figure 5**). Therefore, we concluded that if α-CH_3OH crystals were formed by the heating of a-CH_3OH on a-H_2O grains in young stellar objects (i.e., protoplanetary disks), α-CH_3OH were/are chiral.

Sugisaki et al. [42] observed by calorimetry that the glass transition and crystallization of glassy CH_3OH occurred at about 103 and 105 K, respectively. Luna et al. [14] also observed by IR spectroscopy that the crystallization temperature of a-CH_3OH is between 100 and 110 K.

3.4 NH_3 and its hydrates

3.4.1 NH_3

NH_3 I is a thermodynamically stable phase of solid NH_3 at temperatures <200 K under low pressure [43]. Olovsson and Templeton [24] and Reed and Harris [25] showed by X-ray diffraction and neutron diffraction, respectively, that NH_3 I at 77 K is chiral, including the positions of the hydrogen atoms (space group: $P2_13$). Hewat and Riekel [44] also confirmed this by high-accuracy neutron diffraction at temperatures between 2 and 180 K. It has been observed by IR spectroscopy that the crystallization temperature of a-NH_3 is $\sim$80 K [44, 45].

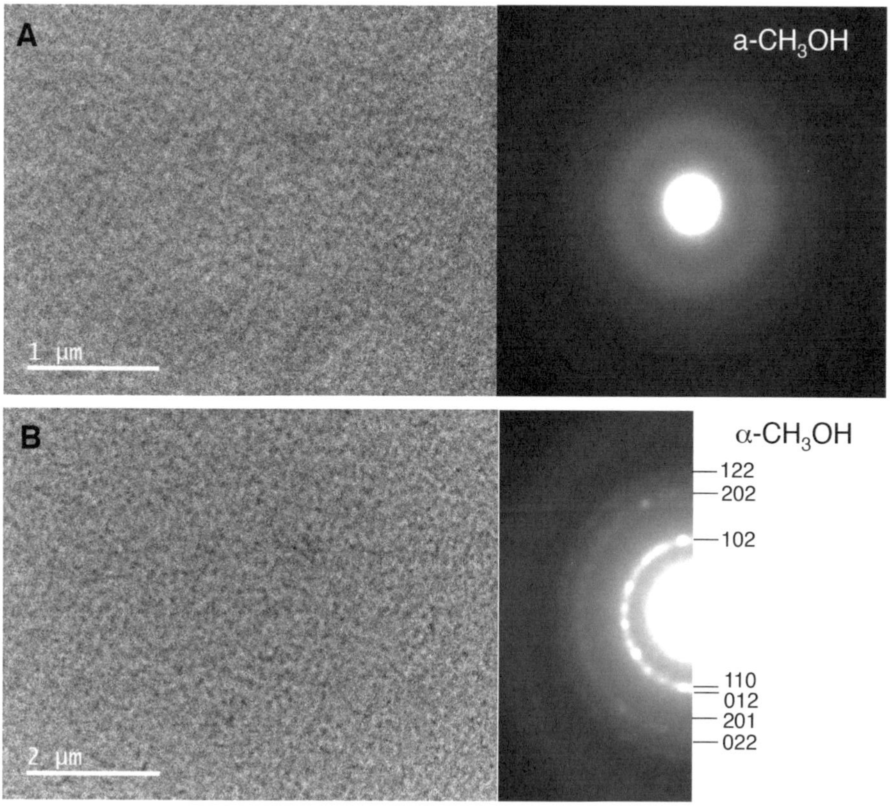

Figure 5.
Transmission electron microscopy image and corresponding electron diffraction pattern of (A) amorphous CH_3OH (a-CH_3OH) deposited on amorphous H_2O (a-H_2O) at 82 K and (B) those of crystalline CH_3OH (α-CH_3OH) at 110 K formed during the heating of a-CH_3OH.

3.4.2 $NH_3{\cdot}H_2O$

$NH_3{\cdot}H_2O$ I is a thermodynamically stable phase of solid $NH_3{\cdot}H_2O$ at temperatures <194 K under low pressure [46]. Olovsson and Templeton [24] showed by X-ray diffraction that $NH_3{\cdot}H_2O$ I at 113 K is chiral, including the positions of the hydrogen atoms (space group: $P2_12_12_1$). Loveday and Nelmes [26] confirmed the space group by neutron diffraction at 110 K.

3.4.3 $NH_3{\cdot}2H_2O$

$NH_3{\cdot}2H_2O$ I is a thermodynamically stable phase of solid $NH_3{\cdot}2H_2O$ at temperatures <176 K under low pressure [47]. Bertie and Shehata [17] showed by X-ray diffraction and IR spectroscopy that $NH_3{\cdot}2H_2O$ I at 100 K is chiral, including the positions of the hydrogen atoms (space group: $P2_12_12_1$). However, neutron diffraction studies showed that the space group is $P2_13$ and the hydrogen is partially ordered at temperatures between 4 and 174 K [26, 27]. Among the four hydrogen sites, the hydrogen occupancy of two of the sites was unity (order), while those of the other two sites were one-third and two-thirds. They considered that the transition to the ordered phase is frustrated by kinetics, as in the transition of pure ice Ih to ice XI.

Fortes et al. [27] suggested the occurrence of a hydrogen-ordered phase (space group: $P2_12_12_1$) at temperatures <140 K because the ordered phase must be thermodynamically more stable than the disordered phase at low temperatures. We therefore suggested the occurrence of a hydrogen-ordered phase (space group: $P2_12_12_1$) at lower temperatures in space, although the equilibrium structure at low temperatures still remains unclear. Further studies on the formation of the ordered phase using a dopant should be undertaken.

3.5 CH_4

CH_4 II is a thermodynamically stable phase of solid CH_4 at temperatures <20.4 K under low pressure [48]. Savoie and Fourier [28] suggested that the CH_4 II space group is $P2_13$ based on the measurement of far-IR spectra at 12 and 25 K. Hashimoto et al. [29] also suggested the same space group based on the calculation of pair-interaction potentials. Press [30] showed by neutron diffraction at 24.5 K that six of the eight molecules were ordered while the remaining two were orientationally disordered, with a space group of $Fm\bar{3}c$. Kobashi et al. [49], based on IR and Raman spectra, suggested that, theoretically, the space group of CH_4 II is $Fm\bar{3}c$. Greiger et al. [50] analyzed the total neutron cross section assuming the structure proposed by Press [30]. Although recent studies have only referred to $Fm\bar{3}c$, these results do not rule out the existence of the space group $P2_13$. Therefore, we cannot eliminate the possibility that both space groups exist.

4. The formation of homochiral ice crystals

4.1 Sources of circularly polarized light

One-handed circularly polarized light (CPL) from an astronomical source could play an essential role in the homochirality of ice crystals. Neutron stars have been suggested as possible sources of CPL [51]; however, CPL at visible and UV wavelengths has not been observed [52], and it is unlikely that a neutron star could encounter a molecular cloud where our solar system was born [53]. In contrast, CPL produced in star-forming regions is considered to be more important because CPL has been observed [54], and the possibility of a star-forming region and a molecular cloud occurring together is very large. Therefore, our discussion of the homochirality of ice crystals assumes that the CPL originated in a star-forming region.

4.2 CPL flux in a molecular cloud

We estimated the CPL photon flux in a molecular cloud based on a simplified model. We assumed two cases: i) a molecular cloud illuminated by the interstellar radiation field and ii) a molecular cloud illuminated by radiation from a nearby massive star. Case i) assumed an isolated star formation, while case ii) assumed a clustered star formation in a massive star-forming region. In both cases, we assumed a 0.1 pc diameter molecular cloud with a hydrogen density of 2×10^5 cm^{-3}. We used Weingartner and Draine's [55] standard dust extinction curve with an R_V parameter of 5.5 to mimic dust in dense clouds [56].

For i), a standard interstellar radiation field model [57] was assumed for the incident radiation spectrum. For (ii), the incident radiation field was simulated by blackbody radiation from a B3-type star (mass = 8 solar mass, luminosity = 2.8 x 10^3 solar luminosity, and surface effective temperature = 2.3 x 10^4 K), which was located 0.1 pc away from the molecular cloud. IR observations have indicated that circularly polarized IR emissions with a degree of circular polarization of up to 20% extend in a 0.1–0.7 pc area in high–/intermediate-mass star-forming regions [58, 59].

We assumed that the CPL was generated within the molecular cloud by the dichroic extinction of incident radiation [60]. A theoretical study predicted that dichroic extinction can produce a degree of circular polarization of up ~10% in star-forming clouds [61]. Here, we assumed that the radiation penetrating the molecular cloud resulted in a 10% degree of circular polarization.

The estimated flux of the CPL in the molecular cloud is summarized in **Figure 6**. On the surface of the molecular cloud, the photon flux reflects sources of radiation and does not change with wavelength. At the middle points (r = 0.025 pc), however, the photon flux decreases with decreasing wavelength. The intensities of the photon fluxes at 200 nm in the cases of i) and ii) were ~ 10^{-1} and ~ 10^3 photons cm^{-2} s^{-1}, respectively, suggesting that the photon flux of case i) was too weak for a photochemical reaction but that of case ii) was effective.

We noted that cosmic-ray-induced UV (CRUV) is a dominant source of UV photons in well-shielded regions [62, 63]. The total photon flux of CRUV is estimated to be 10^4 photons cm^{-2} s^{-1} [64], which is orders of magnitude higher than the estimated photon fluxes at the middle and core points in cases i) and ii). However, because

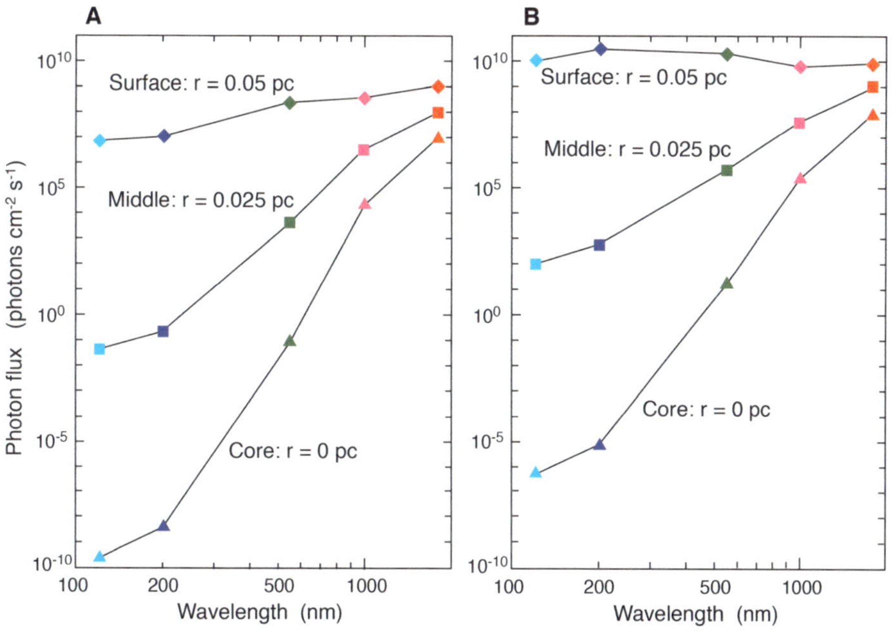

Figure 6.
The wavelength dependence of the flux of circularly polarized light in the molecular cloud. For incident radiation sources, (A) and (B) assume an interstellar radiation field and blackbody radiation from a massive star, respectively. The respective colored marks represent the photon fluxes integrated over the following wavelength regions: Cyan, 90–150 nm; blue, 150–250 nm; green, 400–700 nm; magenta, 800–1200 nm; and red, 1000–2600 nm.

CRUV photons are produced in dense regions, they would be irradiated to icy grains before experiencing dichroic extinction. Thus, we did not consider CRUV to be a source of CPL. If circularly polarized UV light plays an important role in the production of enantiomeric excess, then relevant photo processing would occur on the shallow molecular cloud surface, where the external UV overwhelms the CRUV. Because the volume fraction of the middle part of the molecular cloud is ~0.88, we discuss the asymmetric nucleation of ice crystals using a curve at the middle points (r = 0.025 pc) in the following section.

4.3 Asymmetric nucleation by one-handed CPL

Solid CO is formed via CO deposition from the vapor phase in molecular clouds. The crystallinity of solid CO, either amorphous or crystalline, can be determined by the CO flux in a molecular cloud [5]. Because the CO flux in a molecular cloud is much smaller than the critical flux in which amorphous CO (a-CO) is formed, α-CO should be formed. When one-handed CPL is irradiated during the nucleation of α-CO, the formed crystals might have an enantiomeric excess (**Figure** 7). When there are no metal or high-index nanoparticles on icy grains, α-CO can absorb UV-CPL, which may result in excess enantiomeric crystals. In this case, the formation of one-handed α-CO would only occur in the shallow part of the molecular cloud because the UV-ray penetration depth is not so large (see **Figure 6**). However, when there are metal or high-index nanoparticles on the icy grains, the peak absorption wavelength could be transferred to the visible wavelength region, and the peak could be enhanced compared to that of the UV region [65–67], resulting in excess enantiomeric crystals, possibly up to several tens of percent. This could be supported by laboratory experiments on chiral crystallization [68, 69] and theoretical work [70–72]. In this case, the

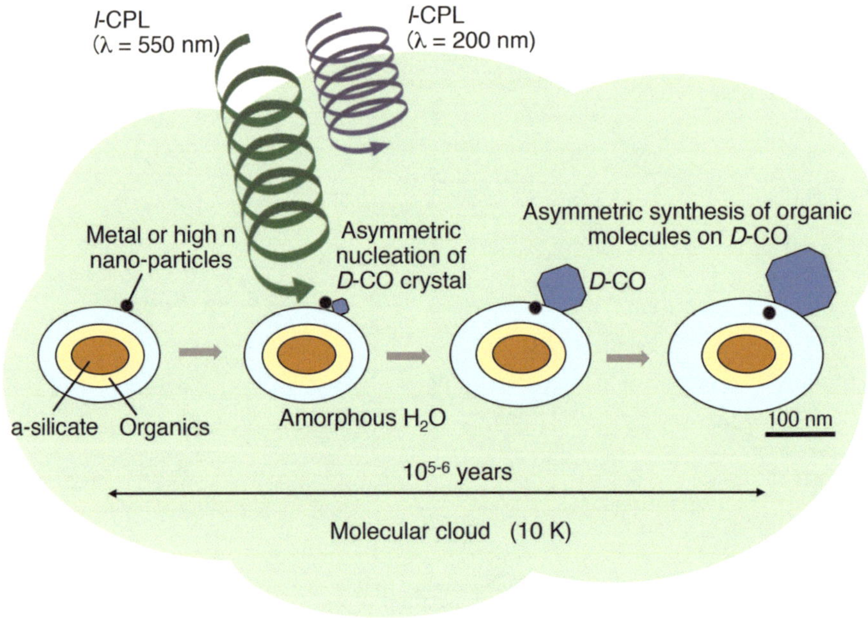

Figure 7.
Schematic illustration of the formation of homochiral CO crystals on the icy grains in molecular clouds.

formation of one-handed α-CO would occur not only in shallow parts of a molecular cloud but also in deeper parts because the penetration depth of visible rays is considerably deep (see **Figure 6**).

A similar process might occur during the crystallization of a-H_2O to form chiral ice crystals and hydrogen-ordered cubic and hexagonal ices (see **Table 2**) in protosolar nebula, as shown in **Figure 1**. The crystallization temperature of a-H_2O under a 10^5-years' timescale is ~90 K [12]. The penetration depth of UV to visible CPL in protoplanetary disks is smaller than that in molecular clouds. However, icy grains could be moved to the surface of the disk by turbulent motion [73] and irradiated with CPL, resulting in the formation of one-handed, hydrogen-ordered H_2O crystals. When ice crystals were recondensed during the cooling of the solar nebula (**Figure 1**), one-handed, hydrogen-ordered H_2O crystals might be formed by the same mechanism.

The crystallization of a-CH_3OH and a-NH_3 also occurred in the protoplanetary disk. Crystallization temperatures of a-CH_3OH and a-NH_3 under a 10^5-years' time-scale can be estimated from those in the laboratory (a-CH_3OH: ~100 K [14, 42] and a-NH_3: ~80 K [15, 45]) and from the assumption that the slopes of a-CH_3OH and a-NH_3 in the plot of the timescale of crystallization vs. the inverse of the temperature lie between those of H_2O [12] and CO_2 [6]. We found that the crystallization temperatures of a-CH_3OH and a-NH_3 under the 10^5-years' timescale were 40–60 K and 20–40 K, respectively. The formation of one-handed α-CH_3OH, NH_3 I, and NH_3 hydrates might also occur, as in the crystallization of a-H_2O. In this way, various kinds of homochiral ice crystals could be formed in protoplanetary disks.

5. Conclusion and outlook

The results of this study indicate the possibility that there were/are many chiral ice crystals in space and that homochiral ice crystals might form by the irradiation of CPL in the star-forming region. These findings have important implications for the origin of the homochirality of organic molecules in space, and the pursuit of the following three suggested areas of study would further our understanding of this.

The crystallinity of CH_3OH and NH_3 in space and the formation mechanism of α-CH_3OH, NH_3 I, and their hydrates in protoplanetary disks are still unclear. Therefore, astronomical observations of the crystallinity of these ices are highly desirable.

For chemical reactions on icy grains, only a-H_2O ice has been considered as a substrate. The adsorption and subsequent surface diffusion of atoms (H, C, N, and O), small molecules (e.g., CO, CO_2, and H_2CO), and radicals (e.g., OH, HCO, and NH), followed by surface two-body reactions to form larger molecules on a-H_2O at low temperatures have been calculated using astrochemical reaction network models [74]. However, this study indicated the possibility of the growth of single ice crystals on grains. On the surface of α-CO, the adsorption behavior of atoms differs greatly from that on a-H_2O [6]. Therefore, it is expected that atoms, except for C and small molecules/radicals, are not adsorbed on the surface of singe-crystalline H_2O ice I. Instead, larger molecules/radicals diffuse easily on the surface of singe-crystalline H_2O ice I, which leads to the formation of more complex organic molecules. Furthermore, the asymmetric adsorption/synthesis of organic molecules on homochiral ice crystals might also proceed.

The search for enantiomeric surfaces on achiral ice crystals, as investigated in minerals [2, 4], is another important subject that should be explored.

Acknowledgements

Part of this work was supported by a Grant-in-Aid for Scientific Research from the Japanese Ministry of Education, Culture, Sports, Science, and Technology and from the Japan Society for the Promotion of Science.

Conflict of interest

The authors declare no competing financial interests.

Author details

Akira Kouchi[1], Takashi Shimonishi[2], Tomoya Yamazaki[1], Masashi Tsuge[1], Naoki Nakatani[3], Kenji Furuya[4], Hiromasa Niinomi[5], Yasuhiro Oba[1], Tetsuya Hama[6], Hiroyasu Katsuno[1], Naoki Watanabe[1] and Yuki Kimura[1*]

1 Institute of Low Temperature Science, Hokkaido University, Sapporo, Japan

2 Center for Transdisciplinary Research, Niigata University, Niigata, Japan

3 Department of Chemistry, Graduate School of Science and Engineering, Tokyo Metropolitan University, Hachioji, Tokyo, Japan

4 National Astronomical Observatory of Japan, Mitaka, Tokyo, Japan

5 Institute of Multidisciplinary Research for Advanced Materials, Tohoku University, Sendai, Japan

6 Komaba Institute for Science, University of Tokyo, Tokyo, Japan

*Address all correspondence to: ykimura@lowtem.hokudai.ac.jp

References

[1] Hazen RM, Sholl DS. Chiral selection on inorganic crystalline surfaces. Nature Materials. 2003;**2**:367-374. DOI: 10.1038/nmat879

[2] Weissbuch I, Lahav M. Crystalline architectures as templates of relevance to the origins of homochirality. Chemical Reviews. 2011;**111**:3236-3267. DOI: 10.1021/cr1002479

[3] Soai K, Asymmetric autocatalysis. Chiral symmetry breaking and the origins of homochirality of organic molecules. Proceedings of the Japan Academy Series B. 2019;**95**:89-110. DOI: 10.2183/pjab.95.009

[4] Hazen RM. Chiral crystal faces of common rock-forming minerals. In: Palyi G, Zucchi C, Caglioti L, editors. Progress in Biological Chirality. Elsevier: Oxford; 2004. pp. 137-151

[5] Kouchi A, Tsuge M, Hama T, Niinomi H, Nakatani N, Shimonishi T, et al. Formation of chiral CO polyhedral crystals on icy interstellar grains. Monthly Notices of the Royal Astronomical Society. 2021;**505**: 1530-1542. DOI: 10.1093/mnras/stab1173

[6] Kouchi A, Tsuge M, Hama T, Oba Y, Okuzumi S, Sirono S-I, et al. Transmission electron microscopy study of the morphology of ices composed of H_2O, CO_2, and CO on refractory grains. The Astrophysical Journal. 2021;**918**:45. DOI: 10.3847/1538-4357/ac0ae6

[7] Boogert ACA, Gerakines PA, Whittet DCB. Observations of the icy universe. Annual Review of Astronomy and Astrophysics. 2015;**53**:541-581. DOI: 10.1146/annurev-astro-082214-122348

[8] Mumma MJ, Charnley SB. The chemical composition of comets-emerging taxinomies and natal heritage. Annual Review of Astronomy and Astrophysics. 2011;**49**:471-524. DOI: 10.1146/annurev-astro-081309-130811

[9] Watanabe N, Kouchi A. Ice surface reactions: A key to chemical evolution in space. Progress in Surface Science. 2008; **83**:439-489. DOI: 10.1016/j.progsurf.2008.10.001

[10] Hagen W, Tielens AGGM, Greenberg JM. The infrared spectra of amorphous solid water and ice Ic between 10 and 140 K. Chemical Physics. 1981;**56**:367-379. DOI: 10.1016/0301-0104(81)80158-9

[11] Whittet DCB. Observation of molecular ices. In: Millar TJ, Williams DA, editors. Dust and Chemistry in Astronomy. Bristol and Philadelphia: Institute of Physics Publishing; 1993. pp. 9-35

[12] Kouchi A, Yamamoto T, Kozasa T, Kuroda T, Greenberg JM. Conditions for condensation and preservation of amorphous ice and crystallinity of astrophysical ices. Astronomy and Astrophysics. 1994;**290**:1009-1018

[13] He J, Toriello FE, Emitiaz SM, Henning T, Vidali G. Phase transition of interstellar CO ice. The Astrpohysical Journal Letters. 2021;**915**:L23. DOI: 10.3847/2041-8213/ac0a7c

[14] Luna R, Molpeceres G, Ortigoso J, Satorre MA, Domingo M, Maté B. Densities, infrared band strengths, and optical constants of solid methanol. Astronomy and Astrophysics. 2018;**617**: A116. DOI: 10.1051/0004-6361/201833463

[15] Zanchet A, Rodríguez-Lazcano Y, Gálvez Ó, Herrero VJ, Escribano R, Maté B. Optical constants of NH_3 and

NH_3:N_2 amorphous ices in the near-infrared and mid-infrared regions. The Astrophysical Journal. 2013;777:26. DOI: 10.1088/0004-637X/777/1/26

[16] Huston T, Hisatsune IC, Heicklen J. Low-temperature infrared studies of some acid-base reactions. Canadian Journal of Chemistry. 1983;**61**: 2077-2088. DOI: 10.1139/v83-361

[17] Bertie JE, Shehata MR. Ammonia dihydrate: Preparation, x-ray powder diffraction pattern and infrared spectrum of $NH_3{\cdot}2H_2O$ at 100 K. The Journal of Chemical Physics. 1984;**81**: 27-30. DOI: 10.1063/1.447381

[18] Raza Z, Alfè D, Salzmann CG, Klimeš J, Michaelides A, Slater B. Proton ordering in cubic ice and hexagonal ice; a potential new ice phase—XIc. Physical Chemistry Chemical Physics. 2011;**13**: 19788-19795. DOI: 10.1039/c1cp22506e

[19] Geiger P, Dellago C, Macher M, Franchini C, Kresse G, Bernard J, et al. Proton ordering of cubic ice Ic: Spectroscopy and computer simulations. Journal of Physical Chemistry C. 2014;**118**: 10989-10997. DOI: 10.1021/jp500324x

[20] Londono JD, Kuhs WF, Finney JL. Neutron diffraction studies of ices III and IX on under-pressure and recovered samples. The Journal of Chemical Physics. 1993;**98**:4878-4888. DOI: 10.1063/1.464942

[21] Wang L, Sun C, Xu H, Zhang J, Zhao Y, Guo W, et al. Neutron diffraction study of crystal structure and temperature driven molecular reorientation in solid α-CO. AIP Advances. 2020;**10**:045301. DOI: 10.1063/1.5121337

[22] Torrie BH, Weng S-X, Powell BM. Structure of the α-phase of solid methanol. Molecular Physics. 1989; **67**:575-581. DOI: 10.1080/00268978900101291

[23] Furuya K, Hama T, Oba Y, Kouchi A, Watanabe N, Aikawa Y. Diffusion activation energy and desorption activation energy for astrochemicaly relevant species on water ice show no clear relation. The Astrophysical Journal Letters. 2022;**933**:L16. DOI: 10.3847/2041-8213/ac78e9

[24] Olovsson I, Templeton D. X-ray study of solid ammonia. Acta Crystallographica. 1959;**12**:832-836. DOI: 10.1107/S0365110X59002420

[25] Reed JW, Harris PM. Neutron diffraction study of solid deuteroammonia. The Journal of Chemical Physics. 1961;**35**:1730-1737. DOI: 10.1063/1.1732137

[26] Loveday JS, Nelmes RJ. The ammonia hydrates—Model mixed-hydrogenbonded systems. High Pressure Research. 2004;**24**:45-55. DOI: 10.1080/08957950410001661990

[27] Fortes AD, Wood IG, Brodholt JP. A high-resolution neutron powder diffraction study of ammonia dehydrate ($ND_3{\cdot}2D_2O$) phase I. The Journal of Chemical Physics. 2003;**119**: 10806-10813. DOI: 10.1063/1.1619371

[28] Savoie R, Fournier RP. Far-infrared spectra of condensed methane and methane-d_4. Chemical Physics Letters. 1970;**7**:1-3. DOI: 10.1016/0009-2614(70)80232-9

[29] Hashimoto M, Hashimoto M, Isobe T. On the crystal structure of methane in phase II. Bulletin of the Chemical Society of Japan. 1971;**44**: 2272-2274. DOI: 10.1246/bcsj.44.2272

[30] Press W. Structure and phase transitions of solid heavy methane

(CD_4). The Journal of Chemical Physics. 1972;**56**:2597-2609. DOI: 10.1063/1.1677586

[31] Mills RL, Olinger B, Cromer DT. Structures and phase diagram of N_2 and CO to 13 GPa by x-ray diffraction. The Journal of Chemical Physics. 1986;**84**:2837-2845. DOI: 10.1063/1.450310

[32] Vegard I. Struktur und Leuchtfähigkeit von festem Kohlenoxyd. Zeitschrift für Physik. 1930;**61**:185-190. DOI: 10.1007/BF01339658

[33] Petrenko VF, Whitworth RW. Physics of Ice. New York: Oxford University Press; 1999. p. 373. DOI: 10.1093/acprof:oso/9780198518945.001.0001

[34] Tajima Y, Matsuo T, Suga H. Phase transition in KOH-doped hexagonal ice. Nature. 1982;**299**:810-812. DOI: 10.1038/299810a0

[35] Leadbetter AJ, Ward RC, Clark JW, Tucker PA, Matsuo T, Suga H. The equilibrium low-temperature structure of ice. The Journal of Chemical Physics. 1985;**82**:424-428. DOI: 10.1063/1.448763

[36] Kawada S. Dielectric dispersion and phase transition of KOH doped ice. Journal of the Physical Society of Japan. 1972;**32**:1442-1442. DOI: 10.1143/JPSJ.32.1442

[37] Minagawa I. Phase transition of cubic ice Ic. Journal of the Physical Society of Japan. 1985;**54**:1610-1614. DOI: 10.1143/JPSJ.54.1610

[38] Lekner J. Electrostatics of proton arrangements in ice Ic. Physica B: Condensed Matter. 1997;**240**:263-272. DOI: 10.1016/S0921-4526(97)00430-4

[39] Casassa S, Calatayud M, Doll K, Minot C, Pisani C. Proton ordered cubic and hexagonal periodic models of ordinary ice. Chemical Physics Letters. 2005;**409**:110-117. DOI: 10.1016/j.cplett.2005.04.068

[40] Kouchi A, Yamazaki T, Katsuno H, Nada H, Hama T, Kimura Y. Possible formation of hydrogen-ordered cubic ice polymorphs by annealing of pure ice Ic at 100–130 K. Submitted to Journal Physical Chemistry

[41] Cervinka C, Beran GJO. Ab initio prediction of the polymorph phase diagram for crystalline methanol. Chemical Science. 2018;**9**:4622-4629. DOI: 10.1039/c8sc01237g

[42] Sugisaki M, Suga H, Seki S. Calorimetric study of the glassy state. III. Novel type calorimeter for study of glassy state and heat capacity of glassy methanol. Bulletin of the Chemical Society of Japan. 1968;**41**:2586-2591. DOI: 10.1246/bcsj.41.2586

[43] Ninet S, Dachi F. High pressure-high temperature phase diagram of ammonia. The Journal of Chemical Physics. 2008;**128**:154508. DOI: 10.1063/1.2903491

[44] Hewat AW, Riekel C. The crystal structure of deutroammonia between 2 and 180 K by neutron powder profile refinement. Acra Crystallographica. 1979;**A35**:569-571. DOI: 10.1107/S0567739479001340

[45] Dawes A, Mukerji RJ, Davis MP, Holtom PD, Webb SM, Sivaraman B. Morphological study into the temperature dependence of solid ammonia under astrochemical conditions using vacuum ultraviolet and Fourier-tansform infrared spectroscopy. The Journal of Chemical Physics. 2007;**126**:244711. DOI: 10.1063/1.2743426

[46] Fortes AD, Suard E, Lemée-Cailleau M-H, Pickard CJ, Needs RJ. Equation of state and phase transition of deuterated ammonia monohydrate ($ND_3{\cdot}D_2O$) measured by high-resolution neutron powder diffraction up to 500 MPa. The Journal of Chemical Physics. 2009;**131**: 154503. DOI: 10.1063/1.3245858

[47] Fortes AD, Wood IG, Brodholt JP, Vočadlo L. The structure, ordering and equation of state of ammonia dehydrate (nh3·2h2o). Icarus. 2003;**162**:59-73. DOI: 10.1016/S0019-1035(02)00073-8

[48] Yamamoto T, Kataoka Y, Okada K. Theory of phase transitions in solid methanes. X. Centering around phase II in solid CH_4. The Journal of Chemical Physics. 1977;**66**:2701-2730. DOI: 10.1063/1.434218

[49] Kobashi K, Okada K, Yamamoto T. Theory of phase transitions in solid methanes. XI. Infrared and Raman spectra of the ν_3 and ν_4 modes in phase II of solid CH_4. The Journal of Chemical Physics. 1977;**66**:5568-5577. DOI: 10.1063/1.433879

[50] Grieger S, Friedrich H, Guckelsberger K, Scherm R, Press W. The total neutron scattering cross section of solid methane in phase II. The Journal of Chemical Physics. 1998;**109**: 3161-3175. DOI: 10.1063/1.476907

[51] Bonner WA. The origin and amplification of biomolecular chirality. Origin of Life and Evolution of the Biosphere. 1991;**21**:59-111. DOI: 10.1007/BF01809580

[52] Bailey J. Astronomical sources of circularly polarized light and the origin of homochirality. Origin of Life and Evolution of the Biosphere. 2001;**31**:167-183. DOI: 10.1023/A: 1006751425919

[53] Greenberg JM, Kouchi A, Niessen W, Irth H, van Paradijs J, de Groot MS, et al. Interstellar dust, chirality, comets and the origins of life: Life from dead stars? Journal of Biological Physics. 1995;**20**:61-70. DOI: 10.1007/BF00700421

[54] Bailey J, Chrysostomou A, Hough JH, Gledhill TM, McCall A, Clark S, et al. Circular polarization in star-formation regions: Implications for biomolecular homochirality. Science. 1998;**281**:672-674. DOI: 10.1126/science.281.5377.672

[55] Weingartner JC, Draine BT. Dust grain-size distributions and extinction in the milky way, large Magellanic cloud, and small Magellanic cloud. The Astrophysical Journal. 2001;**548**: 296-309. DOI: 10.1086/318651

[56] Whittet DCB, Gerakines PA, Hough JH, Shenoy SS. Interstellar extinction and polarization in the Taurus dark clouds: The optical properties of dust near the diffuse/dense cloud interface. The Astrophysical Journal. 2001;**547**:872-884. DOI: 10.1086/318421

[57] Mathis JS, Mezger PG, Panagia N. Interstellar radiation field and dust temperatures in the diffuse interstellar medium and in giant molecular clouds. Astronomy and Astrophysics. 1983;**128**: 212-229

[58] Fukue T, Tamura M, Kandori R, Kusakabe N, Hough JH, Baily J, et al. Extended high circular polarization in the Orion massive star forming region: Implications for the origin of homochirality in the solar system. Origin of Life and Evolution of the Biosphere. 2010;**40**:335-346. DOI: 10.1007/s11084-010-9206-1

[59] Kwon J, Tamura M, Hough JH, Kusakabe N, Nagata T, Nakajima Y, et al.

Near-infrared circular polarization survey in star-forming regions: Correlations and trends. The Astrophysical Journal Letters. 2014;**795**: L16. DOI: 10.1088/2041-8205/795/L16

[60] Martin PG. Interstellar polarization from a medium with changing grain alignment. The Astrophysical Journal. 1974;**187**:461-472. DOI: 10.1086/152655

[61] Lucas PW, Hough JH, Bailey J, Chrysostomou A, Gledhill TM, McCall A. UV circular polarization in star formation regions: The origin of homochirality? Origin of Life and Evolution of the Biosphere. 2005;**35**: 29-60. DOI: 10.1007/s11084-005-7770-6

[62] Prasad SS, Tarafdar SP. UV radiation field inside dense clouds - its possible existence and chemical implications. The Astrophysical Journal. 1983;**267**: 603-609. DOI: 10.1086/160896

[63] Gredel R, Lepp S, Dalgarno A, Herbst E. Cosmic-ray–induced photodissociation and photoionization rates of interstellar molecules. The Astrophysical Journal. 1989;**347**:289-293. DOI: 10.1086/168117

[64] Shen CJ, Greenberg JM, Schutte WA, van Dishoeck EF. Cosmic ray induced explosive chemical desorption in dense clouds. Astronomy and Astrophysics. 2004;**415**:203-215. DOI: 10.1051/0004-6361:20031669

[65] Zhang H, Govorov AO. Giant circular dichroism of a molecule in a region of strong plasmon resonances between two neighboring gold nanocrystals. Physical Review B. 2013; **87**:075410. DOI: 1098-0121/2013/87(7)/075410(8)

[66] Wang R-Y, Wang P, Liu Y, Zhao W, Zhai D, Hong X, et al. Experimental observation of giant chiroptical amplification of small chiral molecules by gold nanosphere clusters. Journal of Physical Chemistry C. 2014;**118**: 9690-9695. DOI: 10.1021/jp5025813

[67] Vestler D, Ben-Moshe A, Markovich G. Enhancement of circular dichroism of a chiral material by dielectric nanospheres. Journal of Physical Chemistry C. 2019;**123**: 5017-5022. DOI: 10.1021/acs.jpcc.8b10975

[68] Niinomi H, Sugiyama T, Tagawa M, Murayama K, Harada S, Ujihara T. Enantioselective amplification on circularly polarized laser-induced chiral nucleation from a $NaClO_3$ solution containing Ag nanoparticles. CrystEngComm. 2016;**18**:7441-7448. DOI: 10.1039/c6ce01464j

[69] Cheng AC, Niinomi H, Omatsu T, Ishida S, Sasaki K, Sugiyama T. Plasmonic manipulation-controlled chiral crystallization of sodium chlorate. Journal of Physical Chemistry Letters. 2020;**11**:4422-4426. DOI: 10.1021/acs.jpclett.0c01041

[70] Tang Y, Cohen AE. Optical chirality and its interaction with matter. Physical Review Letters. 2010;**104**:163901. DOI: 10.1103/PhysRevLett.104.163901

[71] Liu Y, Zhao W, Ji Y, Wang R-Y, Wu X, Zhang XD. Strong superchiral field in hot spots and its interaction with chiral molecules. Europhysics Letters. 2015;**110**:17008. DOI: 10.1209/0295-5075/110/17008

[72] Ho C-S, García-Extarri A, Zhao Y, Dionne J. Enhancing enantioselective absorption using dielectric nanospheres. ACS Photonics. 2016;**4**:197-203. DOI: 10.1021/acsphotonics.6b00701

[73] Ciesla FJ, Sandford SA. Organic synthesis via irradiation and warming of

ice grains in the solar nebula. Science. 2012;**336**:452-454. DOI: 10.1126/science.1217291

[74] Cuppen HM, Walsh C, Lamberts T, Semenov D, Garrod RT, Penteado EM, et al. Grain surface models and data for astrochemistry. Space Science Reviews. 2017;**212**:1-58. DOI: 10.1007/s11214-016-0319-3

Chapter 8

Chiroptical Studies on Anisotropic Condensed Matter: Principle and Recent Applications of the Generalized-High Accuracy Universal Polarimeter

Toru Asahi, Masahito Tanaka, Kenta Nakagawa, Yukana Terasawa, Kazuhiko Ishikawa, Akifumi Takanabe, Hideko Koshima and Bart Kahr

Abstract

Chiroptics is the study of the changes in circular polarization states of light transmitted through analytes typically dissolved in isotropic solutions. However, experimental challenges have long prevented chiroptical measurements of anisotropic media such as single crystals of low symmetry, liquid crystals, or structured films. The high accuracy universal polarimeter (HAUP) was introduced in 1983 to investigate the differential refraction of left and right circular polarization states, circular birefringence (CB), and even in anisotropic media that are dominated by the differential refraction of orthogonal linear polarization states, linear birefringence (LB). In this century, the HAUP was extended to also measure not only the dispersive optical effects (CB and LB) but also the corresponding dissipative effects, circular dichroism (CD) and linear dichroism (LD), differences in light absorption. The improved device is the generalized-HAUP (G-HAUP). Not only can it deliver all the linear optical properties of dissymmetric, anisotropic, and absorbing media, but it can also do so in the ultraviolet as well as the visible part of the electromagnetic spectrum. In this review, characteristic features of the G-HAUP and its applications to crystals of photomechanical salicylidenephenylethylamines, alanine, benzil, and magneto-optical CeF_3 are described.

Keywords: chiroptical properties, circular birefringence, optical activity, circular dichroism, linear birefringence, linear dichroism, chirality, high accuracy universal polarimeter (HAUP), generalized-HAUP (G-HAUP)

IntechOpen

1. Introduction

Chiroptics is the study of the transmission of circularly polarized light through transparent analytes. Linear chiroptical effects are circular birefringence (CB), more colloquially know as optical rotation (OR) or optical activity (OA), sometimes described by gyration of magneto-optical tensors, and circular dichroism (CD). Besides the proliferation of near-synonymous terms, *chiroptics* is itself a misnomer. While chirality is a necessary condition for measurement of CB and CD of molecules rapidly, randomly reorienting in isotropic media, chirality is not necessary when measurements are extended to anisotropic media such as single crystals. Molecules in solution must be chiral, characterized by a pure rotational group (C_n, D_n, T, O, I), to be optically active, but some achiral crystals can also be optically active in some directions, in particular those belonging to non-enantiomorphous crystallographic point groups (C_s, C_{2v}, S_4, D_{2d}) [1]. Thus, the *chiro* prefix is a linguistic infelicity that arose from the historical difficulty of measuring CB and CD in low symmetry crystals, dominated as they are by LB and LD, which can typically be 10^2–10^4 times larger than the former. Molecules are small compared with the wavelength of radiation used in spectropolarimetric measurements; thus, CB and CD are small; molecules of 1–2 nm barely feel the sense of the twist of the electric field of circularly polarized light with a pitch of *ca*. 500 nm. By comparison, it is easy to detect small changes in polarization in isotropic media in normal incidence, which are otherwise indifferent to linearly polarized light. However, in solids with much larger competing perturbations to the state of polarization, chiroptical effects can fall within the experimental noise.

OA was evidenced by Arago in 1811 when he passed linearly polarized light along the high-symmetry *c*-axis of quartz [2]. However, the determination of the anisotropy of OA by making off-axis measurements was a challenging area of research as late as 1988 [3]. The minor dissymmetry-induced perturbations to the polarization state of light in anisotropic media is of the same order of magnitude as parasitic ellipticities from imperfect samples and the polarimeter optical components. Consequently, we know comparatively little about the orientational dependence of CB in molecules, a considerable hole in the science of molecular chirality. A strategy for filling this hole is to characterize the chiroptical properties and to understand what responses contribute to the orientational averages measured in solution.

The *intensity* of light passing through a polarizer, chiral anisotropic sample, and analyzer, contains all the necessary information, in principle, for extracting CB even in the presence of dominating LB. However, the implementation of this extraction was only achieved after the invention of electrophotometry and stable, high-intensity light sources. In 1983, Kobayashi and Uesu used photon-counting techniques, lasers, and computerized modulation of polarizer and the analyzer orientations (θ and Υ; see **Figure 1a**) to determine OA in crystals for directions off the optic axes. This so-called HAUP (high accuracy universal polarimeter) [4–6] was a watershed that gave researchers some confidence after almost two centuries that OA could be measured in crystals generally.

Curiously, the acronym *HAUP* itself is also a slight misnomer. While the HAUP was *universal* in tackling circular and linear anisotropies simultaneously, it could not be considered truly universal until it could function as a complete polarimeter, thus delivering CD and LD as well. For this reason, the extended HAUP method and its associated optical analysis were developed to deliver all four optical effects simultaneously: CB, LB, CD, and LD [6, 7]. Moreover, generalized-HAUP (G-HAUP) has now been extended into the ultraviolet part of the electromagnetic spectrum [8].

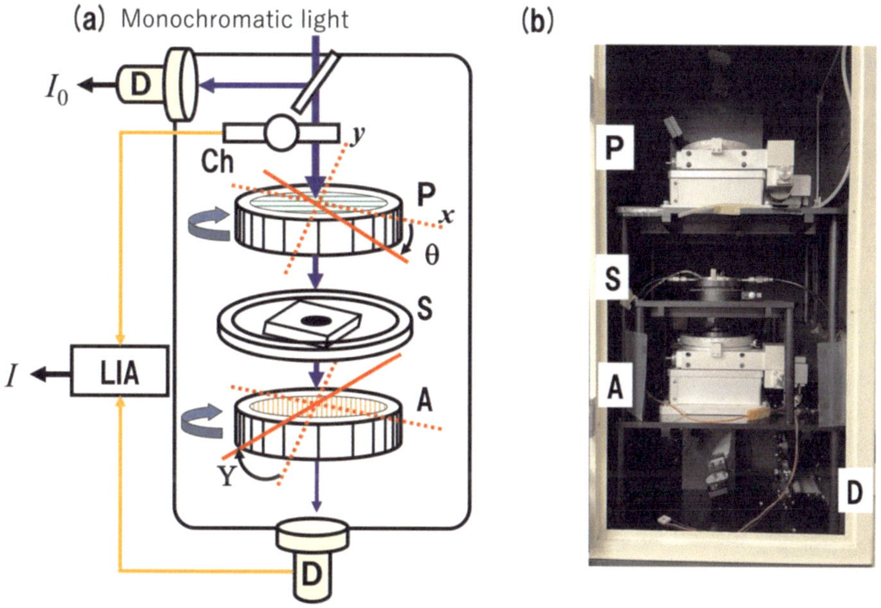

Figure 1.
(a) Schematic representation of the conventional HAUP optics with optical chopper (Ch), lock-in amplifier (LIA), linear polarizer (P), sample stage with a temperature control unit and a pinhole (S), linear analyzer (a), and detector (D). θ and ϒ are the azimuth angle of P from an extinction position of a crystalline sample and the deflecting angle a from the crossed Nicols position, respectively. (b) Photograph of the G-HAUP apparatus developed by authors.

In addition to the HAUP [4–9], other strategies have arisen to measure CB and CD in crystals using modulation of light polarization states photoelastically. These include universal chiroptical spectrophotometer (UCS) [10] and the 4 photo-elastic modulator (4PEM) polarimeter [11]. The UCS relies on the fact that the contributions to a time-varying intensity signal have different frequency dependencies and can be isolated with lock-in amplifiers. 4PEM is a complete polarimeter configured to deliver the whole polarization transfer or Mueller matrix and without any moving optical components to minimize errors. In the 4-PEM polarimeter, all the Mueller matrix elements are simultaneously obtained from 16 frequencies in a Fourier analysis of time-dependent light intensity. The 4-PEM polarimeter can also be operated at oblique angles of incidence so as also to obtain data in reflection like an ellipsometer [12].

This review consists of the following seven sections. Section 2 briefly explains the optical system and characteristic features of the G-HAUP. Sections 3, 4, 5, and 6 describe the recent applications of G-HAUP to salicylidenephenylethylamines [13], alanine [14], benzil [15], and CeF_3 crystals [9]. Finally, Section 7 presents conclusions and future directions for the G-HAUP.

2. Brief description of the G-HAUP method

A schematic representation of the conventional HAUP system is shown in **Figure 1**, which also applies to the G-HAUP. The polarization optics is in both cases composed of two optical elements, a linear polarizer (P) and a linear analyzer (A).

The axes of P and A are set in the crossed Nicols position. The monochromatic light beam emerging from a monochromator or a laser travels through P, the analyte on a sample stage (S), and A successively. Then, the normalized intensity I/I_0 of the light transmitted through P, S, and A is detected with a photomultiplier. θ represents the azimuthal angle of P with respect to a principal axis of the sample, as shown in **Figure 1a**. The positions of P and A are varied systematically from the crossed Nicols position. In addition, Υ is defined as the deflection angle of A from perfect extinction. The transmittance can be expressed using the Jones calculus as a function of a matrix for the polarizer, a generalized Jones matrix containing the linear optical effects propagated through a distance, and a Jones matrix for the analyzer [16]. This expression can be reformulated as a quadratic function of the two variables, θ and Υ angles.

In the HAUP measurement, because the angle between polarization directions of P and A is set close to 90°, I is weak, such that changes are pronounced and can be well determined by an optical chopper and a lock-in amplifier with a photon-counting system.

The HAUP method requires the accurate evaluation and elimination of systematic errors, notably the parasitic ellipticities p and q originating from P and A, respectively, [3, 4] and a small error angle $\delta\Upsilon$ [5]. Here, $\delta\Upsilon$ is caused by the slight deviation from the perfect crossed Nicols condition by inserting the sample between P and A.

In summary, the relative intensity ratio Γ of the transmitted light I to the incident light I_0 is obtained by multiplying the Jones vectors of **P** and **A** and the Jones matrix $\mathbf{M}_\mathrm{H}$ and several approximations and coordinate transformations, as follows:

$$\Gamma(\theta', \Upsilon', \mathrm{LB}, \mathrm{LD}, \mathrm{CB}, \mathrm{CD}, p, q, \delta\Upsilon) = I/I_0 = |\mathbf{A}^\mathrm{T}\mathbf{M}_\mathrm{H}\mathbf{P}|^2 = A'' + B''\Upsilon' + C''\Upsilon'^2 \quad (1)$$

where

$$A'' = H''_{11} + H''_{12}\theta' + H''_{13}\theta'^2.$$
$$B'' = H''_{21} + H''_{22}\theta'.$$
$$C'' = H''_{31}.$$

Each coefficient of the quadratic function of Γ contains the information of θ', Υ', LB, LD, CB, CD, and systematic errors p, q, and $\delta\Upsilon$. Therefore, LB, LD, CB, and CD can be recovered by measuring I at various θ' and Υ' and eliminating the systematic errors.

Here, we would like to point out the requirements of the samples for G-HAUP measurement. As for other optical measurements, transparent, homogeneous, surface-flat, and defect-free samples are preferred for G-HAUP measurements. In addition, although the details are described elsewhere [9], to avoid anomalous behavior near the unstable and low-sensitivity wavelength regions, samples with a small change in the total phase difference with respect to wavelength are also preferred. To fulfill such requirements, we prepared very thin samples for G-HAUP measurements.

The HAUP has been applied to investigate OA of various crystals such as amino acids [17, 18], proteins [19], chiral co-crystals [20], triglycine sulfate (TGS) [21, 22], as well as KH_2PO_4 (KDP) [23] and $NH_4H_2PO_4$ [24] and their isomorphs.

Furthermore, to measure CD and LD in absorbing crystals of low symmetry, the HAUP method has been extended by various researchers [7, 25, 26]. We have developed the extended HAUP method for measuring temperature dependencies of LB, LD, CB, and CD simultaneously in a tris(ethylenediamine) cobalt(III) triiodide monohydrate crystal [27]. The recent applications of G-HAUP are wavelength dependences of chioptical measurements for laminated collagen membranes with highly preferred orientation [28] and crystals of azobenzene-intercalated $K_4Nb_6O_{17}$ [8], γ-glycine [29], salicylidenephenylethylamines [13], *L*-alanine [14], benzil [15], and CeF_3 [9]. In the following sections, analyses of the latter four systems by the G-HAUP are summarized.

3. Chiral photomechanical crystals

The chiral single crystals of *S*- and *R*-enantiomers of *N*-3,5-di-*tert*-butylsalicylidene-1-phenylethylamine in the enol form [enol-(*S*)-**1** and enol-(*R*)-**1**] (**Figure 2**) show photomechanical motion under UV light irradiation [30], that is, light-driven macroscale mechanical motion. Photochromic crystals are often good candidates for photomechanical motion. Upon photoirradiation, the making or breaking of chemical bonds can lead to color changes as well as stresses manifest as macroscale mechanical motion. Photomechanical motion involves the direct conversion of light energy to mechanical energy, and thus, photomechanical crystals may be beneficial for energy conversion. Previous reports show that the dissymmetry of chiral crystals can be manifested in the photomechanical behavior of chiral crystals as opposed to racemic crystals [31, 32].

Enantiomeric salicylidenephenylethylamines enol-(*S*)-**1** and enol-(*R*)-**1** (**Figure 2**) before and under UV light irradiation were analyzed by the G-HAUP method in order

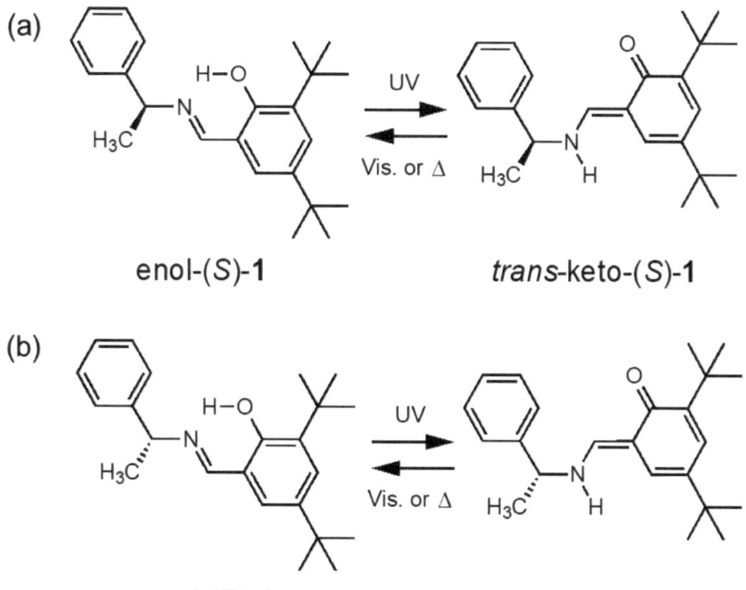

Figure 2.
*Photoinduced hydrogen transfer reaction of salicylidenephenylethylamines enol-(*S*)-1 (a) and enol-(*R*)-1 (b). Reproduced from ref. [13] with permission from the American Chemical Society.*

to obtain LB, LD, CB, and CD spectra and subsequently to correlate the changes of the optical properties to the changes of the crystal structures during the photoreaction [13].

Compounds enol-(*S*)-**1** and enol-(*R*)-**1** were synthesized according to a published protocol, and the single crystals of enol-(*S*)-**1** and enol-(*R*)-**1** were grown by sublimation at 10–20°C below the melting points (92–93°C).

At the outset, the crystal structure of enol-(*S*)-**1** had already been determined, but not that of *trans*-keto-(*S*)-**1** [30]. Hence, *in situ* crystallographic analyses were carefully performed under continuous UV irradiation to analyze the crystal structure of *trans* photoproduct. Because no disorder was found in the crystals, we calculated the crystal structure of *trans*-keto-(*S*)-**1** by dispersion-corrected density functional theory (DFT) calculations (**Figure 3b**). As shown in **Table 1**, the *a* and *b* axes contracted, and the *c* axis extended.

Figure 4 shows the LB, LD, CB, and CD spectra of the enol-(*S*)-**1** (thickness: 6.5 μm) and enol-(*R*)-**1** (thickness: 7.6 μm) crystals through the (001) face before UV irradiation. The LB and LD spectra between the *S* and *R* enantiomeric crystals are coincident (**Figure 4a** and **b**), because linear anisotropies are not affected by enantiomorphism. The negative LD peak at 330 nm corresponds to the π-π* transition of the intramolecularly hydrogen-bonded salicylidenimino moiety. The LB spectra exhibited anomalous dispersion of negative peaks at 360 nm with a change in sign at the strong LD peak. These results show that the LB and LD spectra satisfy the

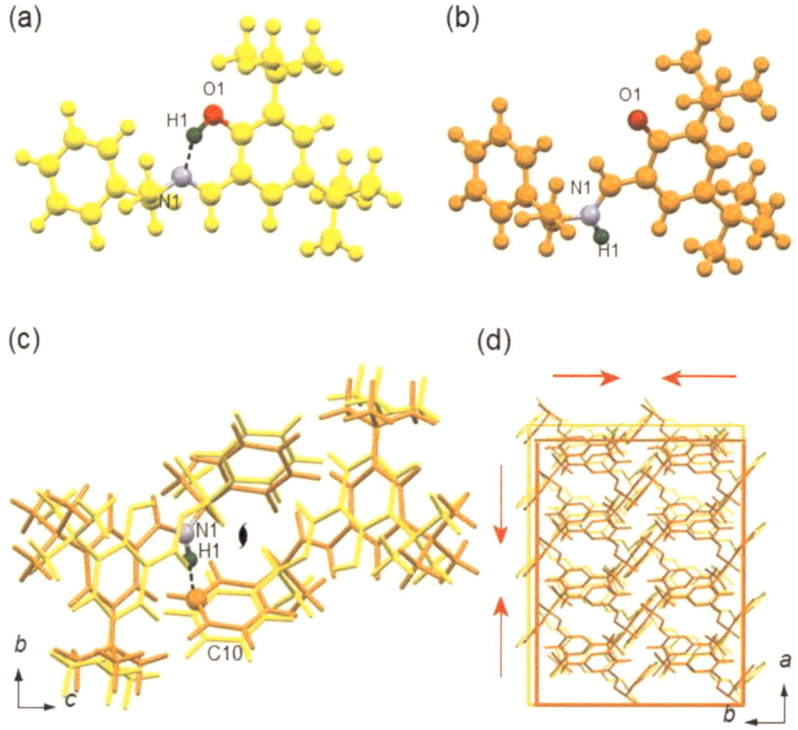

Figure 3.
Calculated crystal structures: Ball-and-stick drawings of (a) enol-(S)-1 (yellow) and (b) trans*-keto-(S)-1 (orange). Tautomer structures overlaid on the (c) (100) and (d) (00$\bar{1}$) faces. The red arrows in (d) show the direction of contraction of the* trans*-keto-(S)-1 crystal along* a *and* b *axes. Reproduced from ref. [13] with permission from the American Chemical Society.*

	enol-(*S*)-1	*trans*-keto-(*S*)-1	Relative change (%)
a (Å)	6.080	5.856	−3.70
b (Å)	9.633	9.313	−3.32
c (Å)	35.454	35.509	+0.16
V (Å^3)	2076.630	1936.486	−6.75

Table 1.
Unit cell dimensions of enol-(S)-1 by X-ray diffraction and trans-keto-(S)-1 crystals obtained from DFT calculations. Reproduced from ref. [13] with permission from the American Chemical Society.

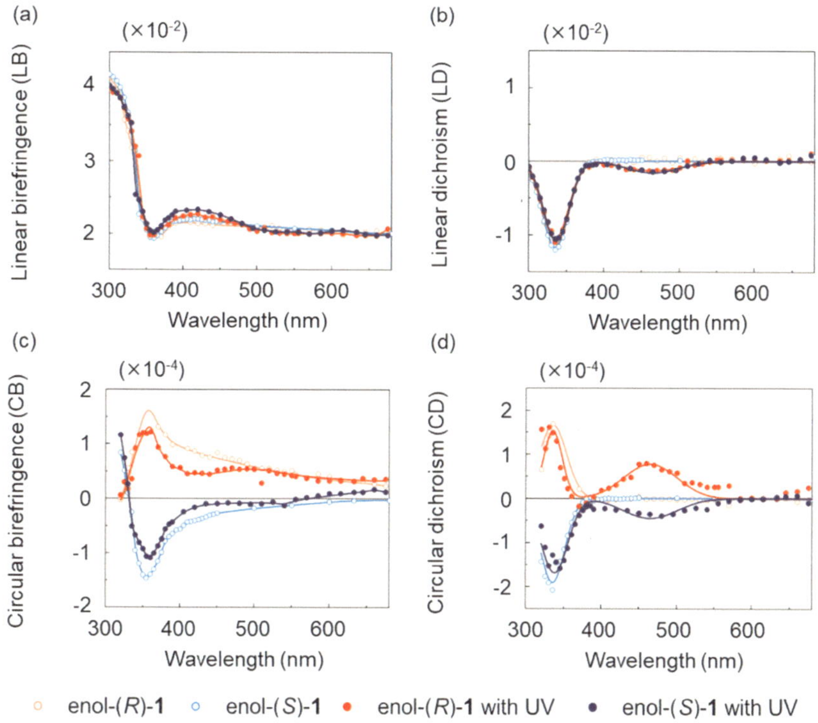

Figure 4.
Optical anisotropic and chiroptical spectra of enol-(S)-1 and enol-(R)-1 crystals on the (001) face: (a) LB, (b) LD, (c) CB, and (d) CD. These properties were measured with the G-HAUP in the dark and under continuous UV light irradiation at 365 nm. The curved lines are fitted gaussian functions intended to guide the eye (a, c). Reproduced from ref. [13] with permission from the American Chemical Society.

Kramers–Kronig relathionship [33]. The CD spectra of enol-(*S*)-**1** and enol-(*R*)-**1** crystals revealed, respectively, a strong negative and positive Cotton effect at 330 nm, which mirror each other (**Figure 4d**). The CB spectra of enol-(*S*)-**1** and enol-(*R*)-**1** crystals also exhibited anomalous dispersion of negative and positive peaks at 360 nm with changes in sign at the CD peaks (**Figure 4c**), respectively, indicating that the Kramers–Kronig relationship also holds between CB and CD.

We then attempted to measure the LB, LD, CB, and CD spectra of both enantiomeric enol-**1** crystals under UV irradiation with a 365 nm LED directed normal to the G-HAUP light path at low power (5 mW cm^{-2}) to minimize incident UV light reaching the detector. UV irradiation-induced bending was inhibited by fixing the crystals to a plate with silicone grease. Note that salicylidenephenylethylamines, which were used in this study, do not show thermochromism. **Figure 4** also shows the LB, LD, CB, and CD spectra under continuous UV irradiation at 365 nm, which represents the spectra at the photostationary state of the reactant and product. New LD peaks corresponding to *trans*-keto-(*S*)-**1** and *trans*-keto-(*R*)-**1** appeared at around 460 nm, and the magnitudes of the LD peaks at 330 nm decreased slightly. The LD spectra of the *S* and *R trans*-keto isomers were coincident (**Figure 4b**). New, small negative and positive CD peaks appeared at 460 nm due to the formation of *trans*-keto-(*S*)-**1** and *trans*-keto-(*R*)-**1** crystals, respectively, and the magnitudes of CD peaks at 330 nm decreased slightly (**Figure 4d**). The CB spectra also exhibited anomalous dispersions of negative [enol-(*S*)-**1**] and positive [enol-(*R*)-**1**] peaks at 500 and 360 nm with a change in sign at the new CD peak (**Figure 4c**), as with the LB and LD spectra. The Kramers–Kronig transformation related the CB to the CD and the LB to the LD.

The LB of both enantiomeric enol-**1** crystals along the *c*-axis showed a relatively lower value (0.02) above 600 nm than the previously reported organic crystals [8, 17–20, 28, 34, 35]. The total intermolecular interaction in the enol-(*S*)-**1** crystal along all directions is the van der Waals force alone. This very weak molecular interaction may have induced such a small LB. In fact, we have reported a much larger value (0.4) for the LB of the chiral cocrystal composed of tryptamine and 4-chlorobenzoic acid [20]. This cocrystal exhibits strong intermolecular interactions, such as ionic bridging and the hydrogen bonding. The optical rotatory power (ORP) value of enol-(*S*)-**1** crystals along the *c*-axis at 632.8 nm before UV irradiation was calculated to be −5.2 deg./mm. The signs of ORP dispersion along the *c*-axis are opposite to those in the hexane solution, the orientationally averaged value [13]. This suggests that the contribution from the ORP dispersion along the *c*-axis might be small, or the ORP dispersions along the *a*- and/or *b*-axes might be largely positive in sign. The dissymmetry parameter, *g*, in the crystalline state is defined as the ratio in absolute magnitude of CD to absorbance. The *g* value of enol-(*S*)-**1** crystal along the *c*-axis was calculated to be 0.013. On the other hand, the *g* value of enol-(*S*)-**1** in hexane solution was calculated to be 0.0010, revealing that the *g* value of the enol-(*S*)-**1** crystal obtained by the G-HAUP measurement without UV irradiation is around 10 times larger than *g* values in the solution and by the calculation in Ref [13].

4. Chiral alanine crystal

L-alanine is the smallest chiral natural amino acid. As an additive in ferroelectric triglycine sulfate crystals, it can control the crystal polarity, which is of practical use in infrared detectors. *L*-alanine alone grows as large, hard, transparent crystals [36] from evaporating solutions in the space group $P2_12_12_1$ [37, 38]. Thus, alanine is an optically biaxial chiral crystal with four zwitterionic molecules ($^+H_3NCH(CH_3)COO^-$) in the unit cell. Misoguti *et al.* [39] measured the dispersion of refractive indices, among the many other physicochemical properties that have been investigated [40–43], but the anisotropy of the optical activity has not been established.

We measured the wavelength dependence of the CB of alanine crystals along each crystallographic axis, by G-HAUP, and assigned the absolute structure of the crystals examined by the method of anomalous dispersion to determine the absolute chirality of alanine crystals, by correlating the absolute structure obtained with the X-ray diffraction method with the CB measured using the G-HAUP.

L- and *D*-alanine crystals were grown by solvent evaporation using the enantiomeric *L*- and *D*-alanine powder as solute and DIW as solvent. Samples for crystal structure determination and the measurement of CB (and optical activity, OA = CB/2) were prepared from these crystals.

The results from the X-ray crystal structure analyses are shown in **Table 2**. The unit cell parameters are almost the same as those from the previous study. Furthermore, we succeeded in determining the absolute structure of the *L*- and *D*-alanine crystals.

For the measurements of chiroptical properties, crystals were cut perpendicular to the <010> direction and (010) faces of *L*- and *D*-alanine crystals were polished to 21 μm and 13 μm, respectively, by using lapping films with SiC (5.0 μm), Al_2O_3 (1.0 μm), and FeO (0.3 μm) abrasive, successively. The polished crystals were fixed on a pinhole and chiroptical properties were measured. The values of LB, LD, and OA in the <010> direction were obtained (**Figure 5**). LB clearly decreases and approaches zero with decreasing wavelength (**Figure 5a**), while LD is almost zero from 280 to 680 nm (**Figure 5b**). LB in the <010> direction is significantly smaller than that in the <100> direction, which allows G-HAUP to detect OA more sensitively. The magnitude of OA is almost the same in *L*- and *D*-alanine crystals, but their signs are opposite as shown in **Figure 5c**. Consequently, *L*- and *D*-alanine crystals exhibit positive and negative OA in the <010> direction, respectively, which means *L*- and *D*-alanine crystals are *levorotatory* and *dextrorotatory*, respectively, in this direction, over the spectrum evaluated.

The interpretation of the results of HAUP studies remains a challenge. In a study of the CB of crystals of *L*-glutamic acid, we introduced the *chirality index*, $r = 1 - |\rho_s|/|\rho_c|$, where $|\rho_s|$ is the absolute optical rotation per molecule in solution and $|\rho_c|$ is the optical rotation per molecule in the crystal averaged over the eigenvalues of the gyration tensor. Values close to 1 are dominated by the effects of crystallization. For *L*-alanine, $r = 0.999$, indicating that the OA is principally a crystal-optical effect.

No general, quantum chemical methods [44] have yet been implemented in widely distributed electronic structure computing programs for interpreting the chiroptical effects of molecular crystals. However, progress is on the horizon. Linear response theories with periodic boundary conditions are required because the OA of molecules is strongly affected by the environment, as confirmed by experimental and computational

	***L*-alanine**	***D*-alanine**	**Simpson Jr. and Marsh [37]**
Temperature (K)	173	173	298
Crystal system	Orthorhombic	Orthorhombic	Orthorhombic
Space group	$P2_12_12_1$	$P2_12_12_1$	$P2_12_12_1$
a (Å)	5.9753(1)	5.9701(4)	6.032(1)
b (Å)	12.2966(2)	12.2935(7)	12.343(1)
c (Å)	5.7895(1)	5.7933(6)	5.784(1)
Z	4	4	4
R_1	0.0250	0.0297	0.049
wR_2	0.0730	0.0764	—
Flack parameter	−0.16(9)	−0.01(11)	—

Crystal data for L-and D-alanine crystals. Reproduced from ref. [14] with permission from Elsevier.

Table 2.
Crystal data for L- *and* D-*alanine crystals.*

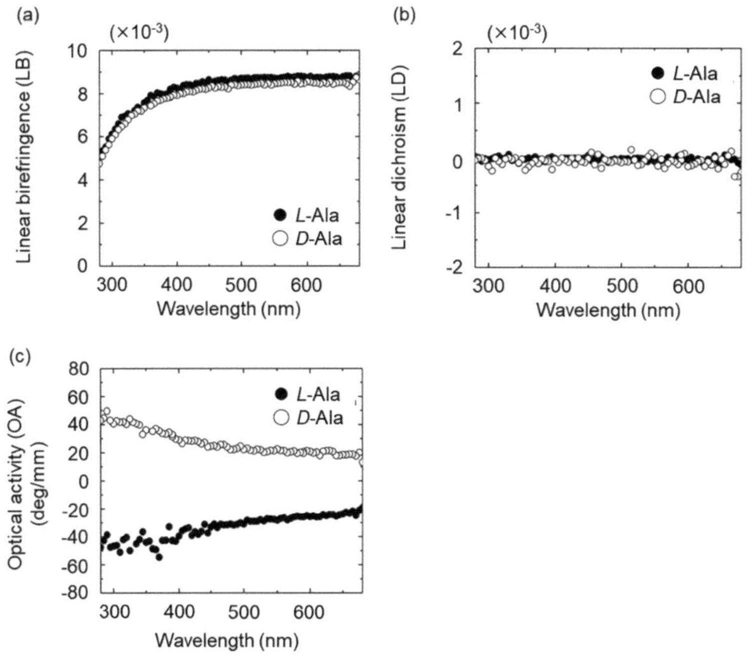

Figure 5.
Wavelength dependence of LB (a), LD (b), and OA (c) in the <010> direction of L- *and* D-*alanine crystals. Closed and opened circles are results for* L- *and* D-*alanine, respectively. Reproduced from ref. [14] with permission from Elsevier.*

studies on the solvent dependence of OR [45], in addition to many studies of computational investigations of crystallographic supercells that we performed over the years. To minimize the effects of interfacial molecules, larger and larger aggregates of molecules must be computed, a process that becomes intractable, as illustrated in the following section for benzil. Balduf and Caricato made this convergence problem explicit in silico for F_2 and HF molecules arranged as model helices [46]. Unit cells were inadequate representations of large helices. More recently, Rérat and Kirtman have introduced computed results of the chiroptical properties of periodic systems using the self-consistent coupled-perturbed method in the program suite CRYSTALS [47].

The relationship between the absolute structure and OA along the *b*-axis was considered parallel to the twofold screw axes considered for *L*-alanine. The hydrogen bond chains for *L*- and *D*-alanine were, respectively, right and left handed (**Figure 6**). However, there is no simple prescription for correlating configuration with the sense of optical rotation. This is true for crystals as for molecules.

From a classical perspective, a right-handed helix of atoms/molecules might be dextrorotatory or levorotatory depending on the polarizability of the groups decorating the helix. From a quantum mechanical perspective, identifying the relevant chromophores is requisite. Furthermore, individual bands may contribute to the CB positively or negatively.

5. Chiral benzil crystal

Benzil ($C_6H_5C(O)$-$C(O)C_6H_5$) crystals have been considered the organic analogue of quartz; both substances have D_3 point symmetry and can be obtained

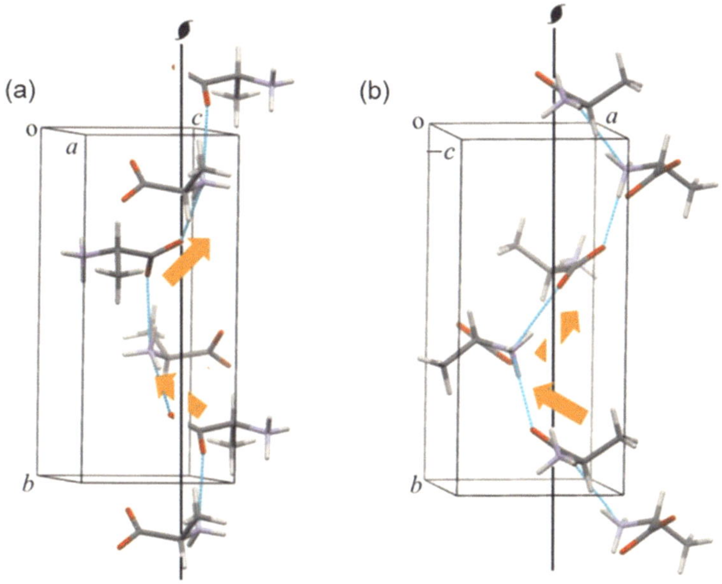

Figure 6.
The handedness of the twofold screw axis of alanine crystals along the b*-axis. (a)* L*-alanine and (b)* D*-alanine. Reproduced from ref. [14] with permission from Elsevier.*

as large single crystals. The OA of benzil has been studied more than that of any other organic crystal; however, unlike quartz, its OA anisotropy has resisted characterization. Without measurements of the optical activity along the diad axes, as opposed to the easily measured optic axis where LB = 0, interpretations are incomplete. Here, we compare OA measurements along the low-symmetry direction of crystalline benzil by the G-HAUP accompanied by electronic structure calculations of the benzil molecule and aggregates of benzil molecules based on the crystal structure.

Single crystals of benzil were grown by slow evaporation from acetone at 25°C. Plates (5 mm × 5 mm × 1 mm) were cut with a razor blade, exposing large (001) or (100) faces. Samples were then polished sequentially with SiC (grain diameter 9 and 5 μm), Al_2O_3 (3 and 1 μm), and Fe_2O_3 (0.3 μm) lapping films. Single-crystal X-ray diffraction analysis confirmed the enantiomorphous space groups $P3_{1(2)}21$ (**Figure 7**). While the structure of benzil has been established previously [48], the absolute structure was determined first in the aforementioned citation.

A *c*-cut slab of $P3_121$ benzil measuring 26.2 μm was analyzed. The OA along the optic axis was easily measured by rotating the analyzer to the extinction position in the G-HAUP. For the $P3_121$ enantiomorph, benzil is *dextrorotatory* at optical frequencies (**Figure 8c**). The OA along the *c*-axis is 24.3°/mm at 590 nm, in good agreement with sodium D-line (25°/mm) measurements [49, 50].

An 88 μm thick *a*-slab of $P3_121$ benzil was likewise polished. The LB, LD, OA, and CD were successfully extracted, and the dispersion in **Figure 8** was fit to a simple Drude oscillator:

$$\rho = \frac{A}{\lambda^2 - \lambda_i^2}, \tag{2}$$

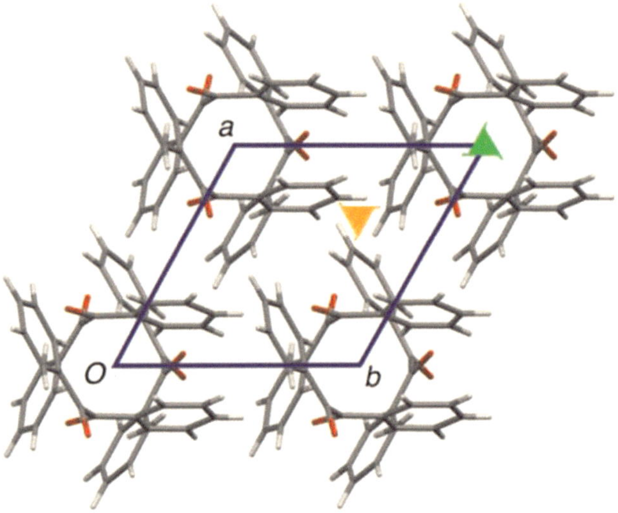

Figure 7.
Unit cell of P$3_1$21 *crystalline benzil viewed along [001] with the 3_1 axes in green and orange. Reproduced from ref. [15] with permission from the American Chemical Society.*

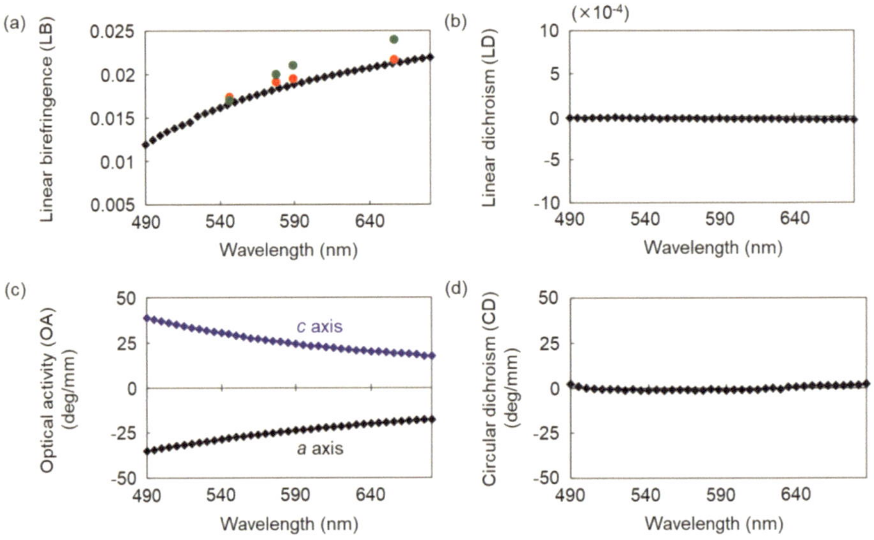

Figure 8.
Wavelength dependences of LB (a), LD (b), fitted OA (c), and CD (d) of the P$3_1$21 *benzil crystal. (a) Published values at fixed wavelengths are red [51] and green [52] points. (c) The OA was fitted to an oscillator model (see text). For the* P$3_1$21 *enantiomorph, the OA along the* c*- and* a*-axes at 590 nm is 24.3 deg./mm and − 23.8 deg./mm, respectively. Reproduced from ref. [15] with permission from the American Chemical Society.*

where A and λ_I are constants with fitted values of 7.9462×10^6 and 117.18 nm, respectively. The LB perpendicular to the optic axis and OA along the optic axis agreed well with literature values [49–52].

The long wavelength (589 nm) OA tensors of one benzil molecule, three benzil molecules in the unit cell, and three benzil molecules related by a threefold rotation

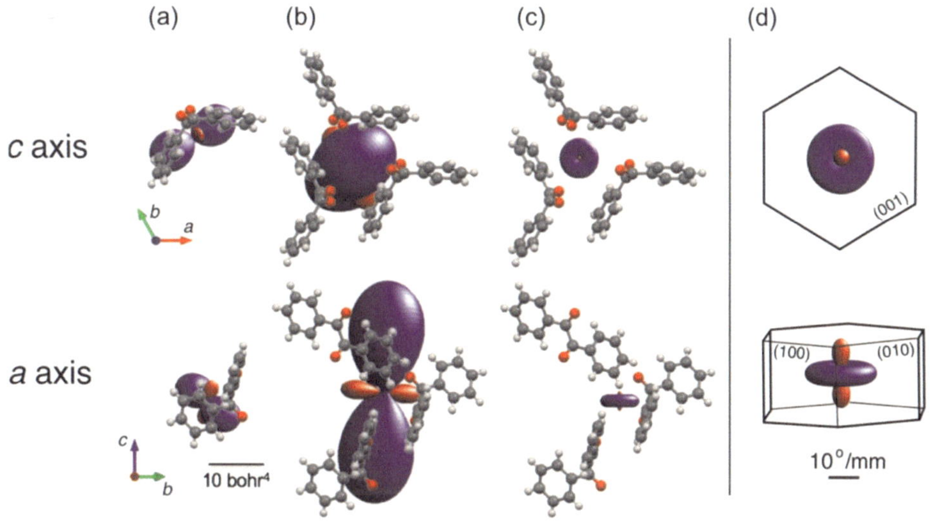

Figure 9.
Representation surfaces of the computed long wavelength OA of benzil at 589 nm (plotted with the software WinTensor, W. Kaminsky). (a) One molecule in the gas phase, (b) one unit cell treated of three molecules, (c) one unit cell based on results for one molecule in (a) and symmetrized, and (d) experimental result based on G-HAUP data. Reproduced from ref. [15] with permission from the American Chemical Society.

(as opposed to a threefold screw) were calculated using well-known methods [53, 54]. The results are summarized in **Figure 9**. The calculations of a small number of molecules are a poor mimic of the crystallographic response.

Benzil and 4-methylbenzophenone [55] are the only such examples of molecular crystals dominated by weak intermolecular interactions for which the long-wavelength OA anisotropy has been determined. Unfortunately, because benzil is in dynamic equilibrium in solution, it is not amenable to a calculation of the chirality index, r (see above).

Interpreting this tensor in terms of a small number of excited states is difficult because unlike simple hydrocarbons investigated previously [56, 57], a great number of states contribute to the long-wavelength value of benzil. Requisite for the computation of chiroptical properties in crystals is the development of linear response theory with periodic boundary conditions to provide a framework for interpreting the results of single-crystal polarimetry, as discussed above.

6. Magneto-optical CeF_3 crystal

Frequency instabilities and parasitic oscillations in optical sources are deleterious to a civilization increasingly dependent on the transfer of optical information with light. An optical isolator is an optical diode that allows light to pass in only one direction. The functions of optical isolators are based on the non-reciprocal magneto-optical rotation of polarization of light, long known as the Faraday rotation (FR). The crystals investigated and employed as optical isolators to date include only isotropic crystals or uniaxial crystals with well-defined isotropic directions (*i.e.*, along the optic axis). Otherwise, optically anisotropic crystals are complicated by LB and LD.

Like OA, FR in anisotropic crystals cannot be accurately measured conventionally, except along high-symmetry directions. Therefore, low-symmetry directions in anisotropic crystals have been avoided in magneto-optical research.

A sample subjected to a magnetic field applied parallel and anti-parallel to the wave vector of the light can manifest FR and magnetic circular dichroism (MCD) that can be analyzed by the G-HAUP method. A Nd-Fe-B (NIB) magnet introduced for this purpose is shown in **Figure 10**. In this configuration, we measured the dispersion of FR and MCD in CeF_3 single crystal along the optic axis (c-axis) as well as *perpendicular* to the optic axis (a-axis) with the G-HAUP.

Single crystals of CeF_3 were grown by the modified Czochralski technique [58–60]. We prepared two types of samples: (001) and (100) plates of single-crystal CeF_3. The smooth flat sample surfaces were obtained by polishing machines with a diamond slurry (3.0 μm) and colloidal silica (32.5 nm) abrasive, successively. We polished a 307 μm thick (001) plate of a single crystal of CeF_3 [58–60] with point symmetry D_{3d}. When the magnetic field was applied parallel to the light propagation direction, FR occurs only along <001>. In this direction, the measurement is conventional, but here, we implemented the rotating analyzer mode of G-HAUP. The sample was sandwiched between two permanent NIB magnets and illuminated through a 0.5 mm pinhole (**Figure 10**). The longitudinal field of ~0.5 Tesla (T) was homogeneous [60]. The wavelength dependence of the Verdet constant, the quantity in deg./mm·T that describes the strength of FR along the c-axis at 25°C, is plotted as black rhombuses in **Figure 11c**. Positive Verdet constants along the c-axis indicated that the right-handed

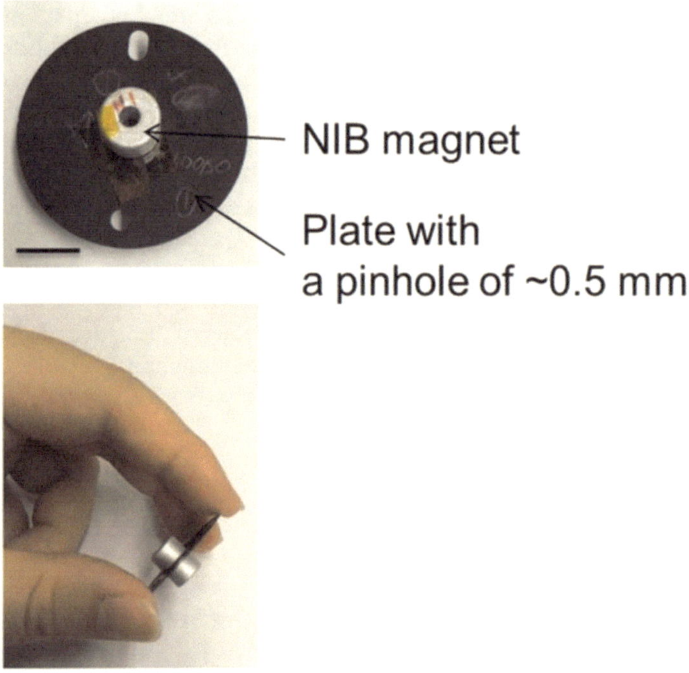

Figure 10.
Photograph of the Nd-Fe-B (NIB) magnet with the magnetic field applied parallel or anti-parallel to the light propagation direction. The sample was mounted on a plate with a ~ 0.5 mm pinhole sandwiched between two identical NIB permanent magnet rings. The black scale bar = 1 cm. Reproduced from ref. [9] with permission from Springer Nature.

circularly polarized light propagates faster than its left-handed counterpart across the spectrum.

A (100) plate just 58.0 μm thick was polished. The extinction directions were determined accurately with an Ehringhaus compensator fitted to a polarized light microscope. We first determined with G-HAUP in the extended mode the dispersion of LD, OA, and CD along the *a*-axis in the absence of the magnetic field at 25°C (blue rhombuses in **Figure 11**). LB was comparable to that of *α*-quartz (**Figure 11a**). LD was almost zero over the wavelength examined (**Figure 11b**), consistent with the absence of absorption bands at wavelengths longer than 282 nm. The OA and CD values were also close to zero, consistent with the D_{3d} symmetric crystal structure. The dispersion of LB, LD, FR, and MCD along <100> was established, while the magnetic field was propagated parallel and anti-parallel to the wave vector at 25°C (red rhombuses in **Figure 11**). LB and LD were indifferent to the sign of the magnetic field. Reversing the magnetic field direction inverted the signs of FR and MCD. The values of LB and LD hardly changed with and without the magnetic field (**Figure 11a** and **b**). The Verdet constants along the *a*-axis were positive throughout the wavelength region (**Figure 11c**).

CeF_3 is paramagnetic at room temperature. Therefore, the magnetic interactions such as ferromagnetism, anti-ferromagnetism, or ferrimagnetism, which are often observed in low-temperature regions, do not affect the G-HAUP measurements.

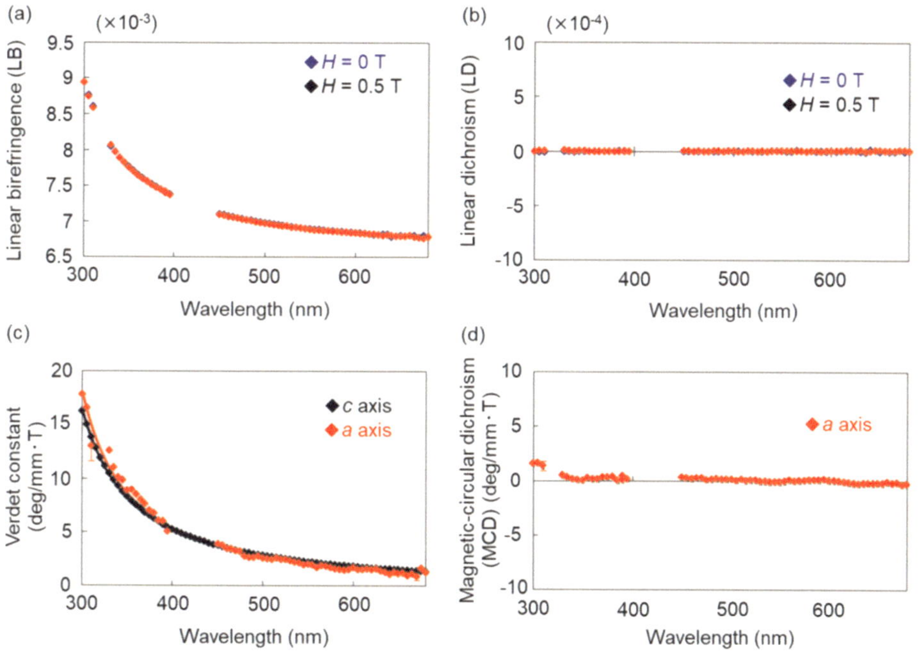

Figure 11.
Wavelength dependences of LB (a), LD (b), Verdet constant (c), and MCD (d) in single-crystal CeF_3 at 25°C. the black, blue, and red rhombuses represent the data along <001>, along <100> without a magnetic field, and along <100> under an applied magnetic field, respectively. We eliminated points in spectral regions where G-HAUP is less sensitive. The Verdet constant spectra were fitted to a simple Drude oscillator. The error bars represent the standard deviations. The error along <100> given in (c) ranged from 0.32 to 1.40 deg./mm·T. reproduced from ref. [9] with permission from Springer Nature.

Cerium is usually in the 3+ oxidation state in condensed matter; the electronic configuration of Ce^{3+} is $1s^2 2s^2 2p^6$ ... $4d^{10} 4f^1 5s^2 5p^6$. Transitions from $4f \rightarrow 5d$ confer the magneto-optical properties in the UV–Vis-IR region [61]. The Verdet constants were positive along the *c*- and *a*-axes at all measured wavelengths, and their magnitudes along both axes were nearly equal (**Figure 11c**). The magnetic susceptibilities, measured with SQUID (superconducting quantum interference device), were nearly identical in both directions. The magnitude of Verdet constants at an arbitrary frequency depends on the magnetic susceptibility and the transition probabilities. Therefore, we may surmise that the anisotropy of the summed $4f \rightarrow 5d$ transition probabilities is zero.

7. Conclusions

In this review, principles and recent applications of G-HAUP method are concisely introduced. New CD peaks in photomechanical crystals originating from photochromism were observed. According to the G-HAUP results of alanine crystals, we discuss the relationship between the absolute structure and OA. In addition, we found by benzil measurements and quantum chemical calculations that the intermolecular interactions are decisive even though comparatively weak dispersion forces dominate the interactions between molecules. Moreover, the first application of G-HAUP to a magneto-optical material was presented by applying magnetic field with NIB magnets. These results strongly indicate the usefulness of the HAUP method for evaluating the chiroptical and magneto-optical properties of ordered specimens. More recently, a rapid HAUP system was developed by using dispersive detection technique with a CCD array spectrometer [62]. Very thin crystalline sample, *ca.* less than several dozen micrometers, were prepared to accommodate the G-HAUP transmission requirement. A reflection-mode HAUP system can contribute to the significant increase in the number of HAUP-applicable specimen because of the simplification of sample preparation. In addition, the extension of wavelength coverage to the infrared and/or vacuum UV region results in the significant increase in the information related to the electronic structure and/or conformation of optically active specimens [63].

Acknowledgements

This review is cordially dedicated to Prof. Jinzo Kobayashi, who is an inventor of HAUP and has opened a new way to solid-state dissymmetry from a spectroscopic point of view. We thank Prof. Dr. Hidehiro Uekusa, Prof. Dr. Tadashi Mori, Dr. Alexander T. Martin, Dr. Shane M. Nichols, and Dr. Veronica L. Murphy for their fruitful discussions and comments. This study was financially supported by the JSPS Scientific Research in the Challenging Exploratory Research, the High-Tech Research Center (TWIns), the Consolidated Research Institute for Advanced Science and Medical Care (ASMeW), the Global COE for Practical Chemical Wisdom, the Leading Graduate Program in Science and Engineering, the Top Global University Project, Waseda University, from the Ministry of Education, Culture, Sports, Science and Technology, Japan, and the grant-in-aid from the Mitsubishi Materials Corporation and the Mizuho Foundation for the Promotion of Sciences.

Conflict of interest

The authors declare no conflict of interest.

Author details

Toru Asahi[1,2*], Masahito Tanaka[3], Kenta Nakagawa[1,4*], Yukana Terasawa[1,4], Kazuhiko Ishikawa[1], Akifumi Takanabe[1], Hideko Koshima[2] and Bart Kahr[5]

1 Graduate School of Advanced Science and Engineering, Waseda University, Tokyo, Japan

2 Research Organization for Nano and Life Innovation, Waseda University, Tokyo, Japan

3 National Institute of Advanced Industrial Science and Technology (AIST), Tokyo, Japan

4 Kagami Memorial Research Institute for Materials Science and Technology, Waseda University, Tokyo, Japan

5 Department of Chemistry and Molecular Design Institute, New York University, New York, USA

*Address all correspondence to: tasahi@waseda.jp
and kenta.nakagawa@aoni.waseda.jp

Rcferences

[1] Claborn K, Isborn C, Kaminsky W, Kahr B. Optical rotation of achiral compounds. Angewandte Chemie (International Ed. in English). 2008;**47**: 5706-5717. DOI: 10.1126/science.1173605

[2] Kahr B, Arteaga O. Arago's best paper. ChemPhysChem. 2012;**13**:79-88. DOI: 10.1002/cphc.201100660

[3] Kobayashi J, Asahi T, Takahashi S, Glazer AM. Evaluation of the systematic errors of polarimetric measurements: Application to measurements of the gyration tensors of α-quartz by the HAUP. Journal of Applied Crystallography. 1988;**21**:479-484

[4] Kobayashi J, Uesu Y. A new optical method and apparatus 'HAUP' for measuring simultaneously optical activity and birefringence of crystals. I. Principles and construction. Journal of Applied Crystal. 1983;**16**:204-211. DOI: 10.1107/S0021889883010262

[5] Kobayashi J, Kumomi H, Saito K. Improvement of the accuracy of HAUP, high-accuracy universal polarimeter: Application to ferroelectric $[N(CH_3)_4]ZnCl_4$. Journal of Applied Crystallography. 1986;**19**:377-381. DOI: 10.1107/S0021889886089197

[6] Asahi T, Kobayashi J. Polarimeter for anisotropic optically active materials. In: Weiglhofer WS, Lakhtakia A, editors. Introduction to Complex Mediums for Optics and Electromagnetics. Washington: SPIE; 2003. pp. 645-676. DOI: 10.1117/3.504610.ch26

[7] Kobayashi J, Asahi T, Sakurai M, Takahashi M, Okubo K, Enomoto Y. Optical properties of superconducting $Bi_2Sr_2CaCu_2O_8$. Physical Review B. 1996; **53**:11784-11795. DOI: 10.1103/PhysRevB. 53.11784

[8] Tanaka M, Nakamura N, Koshima H, Asahi T. An application of the advanced high-accuracy universal polarimeter to the chiroptical measurement of an intercalated compound $K_4Nb_6O_{17}$ with high anisotropy. Journal of Physics D: Applied Physics. 2012;**45**:175303. DOI: 10.1088/0022-3727/45/17/175303

[9] Nakagawa K, Asahi T. Determination of the faraday rotation perpendicular to the optical axis in uniaxial CeF_3 crystal by using the generalized-high accuracy universal polarimeter. Scientific Reports. 2019;**9**:18453. DOI: 10.1038/s41598-019-54174-2

[10] Kuroda R, Harada T, Shindo Y. A solid-state dedicated circular dichroism spectrophotometer: Development and application. The Review of Scientific Instruments. 2001;**72**:3802-3810. DOI: 10.1063/1.1400157

[11] Arteaga O, Freudenthal J, Wang B, Kahr B. Mueller matrix polarimetry with four photoelastic modulators: Theory and calibration. Applied Optics. 2012; **51**:6805-6817. DOI: 10.1364/AO.51. 006805

[12] Arteaga O, Freudenthal J, Kahr B. Reckoning electromagnetic principles with polarimetric measurements of anisotropic optically active crystals. Journal of Applied Crystallography. 2012;**45**:279-291. DOI: 10.1107/S002188 9812006085

[13] Takanabe A, Tanaka M, Johmoto K, Uekusa H, Mori T, Koshima H, et al. Optical activity and optical anisotropy in photomechanical crystals of chiral salicylidenephenylethylamines. Journal of the American Chemical Society. 2016; **138**:15066-15077. DOI: 10.1021/ jacs.6b09633

[14] Ishikawa K, Terasawa Y, Tanaka M, Asahi T. Accurate measurement of the optical activity of alanine crystals and the determination of their absolute chirality. Journal of Physics and Chemistry of Solids. 2017;**104**:257-266. DOI: 10.1016/j.jpcs.2017.01.024

[15] Nakagawa K, Martin AT, Nichols SM, Murphy VL, Kahr B, Asahi T. Optical activity anisotropy of benzil. Journal of Physical Chemistry C. 2017;**121**:25494-25502. DOI: 10.1021/acs.jpcc.7b08831

[16] Jones RC. A new calculus for the treatment of optical systems. VII. Properties of the N-matrices. Journal of the Optical Society of America. 1948;**38**: 671-685. DOI: 10.1364/JOSA.38.000671

[17] Asahi T, Utsumi H, Itagaki Y, Kagomiya I, Kobayashi J. Optical activity of crystalline glutamic acids. Acta Crystallography. 1996;**A52**:766-769. DOI: 10.1107/S010876739609993X

[18] Asahi T, Takahashi M, Kobayashi J. The optical activity of crystalline *L*-aspartic acid. Acta Crystallography. 1997;**A53**:763-771. DOI: 10.1107/S0108767397004595

[19] Kobayashi J, Asahi T, Sakurai M, Kagomiya I, Asai H, Asami H. The optical activity of lysozyme crystals. Acta Crystallography. 1998;**54**:581-590. DOI: 10.1107/s0108767398001986

[20] Koshima H, Nagano M, Asahi T. Optical activity induced by helical arrangements of tryptamine and 4-chlorobenzoic acid in their cocrystal. Journal of the American Chemical Society. 2005;**127**:2455-2463. DOI: 10.1021/ja044472f

[21] Herreros-Cedres J, Hernandez-Rodriguez C, Castro MP, Montoya MM, Rodriguez VL. Optical anisotropy of $TGS_x \cdot TGSe_{1-x}$ compounds. Ferroelectrics. 2007;**350**:22-28. DOI: 10.1080/00150190701369602

[22] Kaminsky W. Experimental and phenomenological aspects of circular birefringence and related properties in transparent crystals. Reports on Progress in Physics. 2000;**63**:1575-1640. DOI: 10.1088/0034-4885/63/10/201

[23] Shopa Y, Lutsiv-Shumskiy L, Serkiz R. Optical activity of the KDP group crystals. Ferroelectrics. 2005;**317**: 79-82. DOI: 10.1080/00150190590963480

[24] Kaminsky W, Steininger S, Herreros-Cedres J, Glazer AM. Evidence of a circularly polarized light mode along the optic axis in *c*-cut $NH_4H_2PO_4$, induced by circular differential reflection and anomalous birefringence. Journal of Physics: Condensed Matter. 2010;**22**:095902. DOI: 10.1088/0953-8984/22/9/095902

[25] Moxon JRL, Remshaw AR. The simultaneous measurement of optical activity and circular dichroism in birefringent linearly dichroic crystal sections. I. Introduction and description of the method. Journal of Physics. 1990; **2**:6807-6836. DOI: 10.1088/0953-8984/2/32/012

[26] Kaminsky W, Glazer AM. Measurement of optical rotation in crystals. Ferroelectrics. 1996;**183**: 133-141. DOI: 10.1080/00150199608224099

[27] Matsuki R, Asahi T, Kobayashi J, Asai H. Measurement of circular birefringence and circular dichroism of the single crystals of Λ-$(+)_{589}$- and Δ-$(-)_{589}$-tris(ethylenediamine)cobalt (III) triiodide monohydrate by the extended HAUP method. Chirality.

2004;**16**:286-293. DOI: 10.1002/chir.20029

[28] Nakagawa K, Harper-Lovelady H, Tanaka Y, Tanaka M, Yamato M, Asahi T. A high-accuracy universal polarimeter study of optical anisotropy and optical activity in laminated collagen membranes. Chemical Communications. 2014;**50**:15086-15089. DOI: 10.1039/C3CC49328H

[29] Ishikawa K, Tanaka M, Suzuki T, Sekine A, Kawasaki T, Soai K, et al. Absolute chirality of the γ-polymorph of glycine: Correlation of the absolute structure with the optical rotation. Chemical Communication. 2012;**48**: 6031-6033. DOI: 10.1039/C2CC30549F

[30] Koshima H, Matsuo R, Matsudomi M, Uemura Y, Shiro M. Light-driven bending crystals of salicylidenephenylethylamines in enantiomeric and racemate forms. Crystal Growth & Design. 2013;**13**: 4330-4337. DOI: 10.1021/cg400675r

[31] Kitagawa D, Nishi H, Kobatake S. Photoinduced twisting of a photochromic diarylethene crystal. Angewandte Chemie, International Edition. 2013;**52**:9320-9322. DOI: 10.1002/anie.201304670

[32] Taniguchi T, Fujisawa J, Shiro M, Koshima H, Asahi T. Mechanical motion of chiral azobenzene crystals with twisting upon photoirradiation. Chemistry - A European Journal. 2016; **22**:7950-7958. DOI: 10.1002/chem.201505149

[33] Toll JS. Causality and the dispersion relation: Logical foundations. Physics Review. 1956;**104**:1760-1770. DOI: 10.1103/PhysRev.104.1760

[34] Kobayashi J, Uchino K, Asahi T. Optical properties of Rochelle salt. Physical Review B. 1991;**43**:5706-5712. DOI: 10.1103/PhysRevB.43.5706

[35] Asahi T, Nakamura M, Kobayashi J, Toda F, Miyamoto H. Optical activity of oxo amide crystals. Journal of the American Chemical Society. 1997;**119**: 3665-3669. DOI: 10.1021/ja9620189

[36] Raj JC, Dinakaran S, Krishnan S, Boaz MB, Robert R, Das JS. Studies on optical, mechanical and transport properties of NLO active *L*-alanine formate single crystal grown by modified Sankaranarayanan–Ramasamy (SR) method. Optics Communication. 2008;**281**:2285-2290. DOI: 10.1016/j.optcom.2007.12.019

[37] Simpson HJ Jr, Marsh RE. The crystal structure of *L*-alanine. Acta Crystallography. 1966;**20**:550-555. DOI: 10.1107/S0365110X66001221

[38] Lehmann MS, Koetzle TF, Hamilton WC. Precision neutron diffraction structure determination of protein and nucleic acid components. I. Crystal and molecular structure of the amino acid L-alanine. Journal of the American Chemical Society. 1972;**94**: 2657-2660. DOI: 10.1021/ja00763a016

[39] Misoguti L, Valera AT, Nunes FD, Bagnato VS, Melo FEA, Mendes FJ, et al. Optical properties of *L*-alanine organic crystals. Optical Materials. 1996;**6**: 147-152. DOI: 10.1016/0925-3467(96)00032-8

[40] Razzetti C, Ardoino M, Zanotti L, Zha M, Paorici C. Solution growth and characterisation of *L*-alanine single crystals. Crystal Research and Technology. 2002;**37**:456-465. DOI: 10.1002/1521-4079(200205)37:5<456::AID-CRAT456>3.0.CO;2-M

[41] Wojciechowski A, Ozga K, Reshak AH, Miedzinski R, Kityk IV,

Berdowski J, et al. Photoinduced effects in *L*-alanine crystals. Materials Letters. 2010;**64**:1957-1959. DOI: 10.1016/j.matlet.2010.06.034

[42] Bisker-Leib V, Doherty MF. Modeling crystal shape of polar organic materials: Applications to amino acids. Crystal Growth & Design. 2003;**3**: 221-237. DOI: 10.1021/cg025538q

[43] Vijayan N, Rajasekaran S, Bhagavannarayana G, Ramesh BR, Gopalakrishnan R, Palanichamy M, et al. Growth and characterization of nonlinear optical amino acid single crystal: *L*-alanine. Crystal Growth & Design. 2006;**6**:2441-2445. DOI: 10.1021/cg049594y

[44] Zhong H, Levine ZH, Allan DC, Wilkins JW. Optical activity of selenium: A nearly first-principles calculation. Physical Review Letters. 1992;**69**: 379-382. DOI: 10.1103/PhysRevLett.69.379

[45] Lahiri P, Wiberg KB, Vaccaro PH, Caricato M, Crawford TD. Large solvation effect in the optical rotatory dispersion of norbornenone. Angewandte Chemie. 2014;**126**: 1410-1413. DOI: 10.1002/anie.201306339

[46] Balduf T, Caricato M. Helical chains of diatomic molecules as a model for solid-state optical rotation. Journal of Physical Chemistry C. 2019;**123**: 4329-4340. DOI: 10.1021/acs.jpcc.8b12084

[47] Rérat M, Kirtman B. First-prinnciples calculation of the optical rotatory power of periodic systems: Application on α-quartz, tartaric acid crystal, and chiral (n,m)-carbon nanotubes. Journal of Chemical Theory and Computation. 2021;**17**:4063-4076. DOI: 10.1021/acs.jctc.1c00243

[48] More M, Odou G, Lefebvre J. Structure determination of benzil in its two phases. Acta Crystallography. 1987; **43**:398-405. DOI: 10.1107/S0108768187097660

[49] Chandrasekhar S. The rotatory dispersion of benzil. Proceedings of the Indian Academy Science. 1954;**39**:243-253. DOI: 10.1007/BF03047144

[50] Chaudhuri NK, El-Sayed MA. Molecular origin of the optical rotatory dispersion of the benzil crystal. The Journal of Chemical Physics. 1967;**47**: 1133-1143. DOI: 10.1063/1.1711998

[51] Jelley E. Application of grating microspectrograph to problem of identifying organic compounds. Industrial and Engineering Chemistry, Analytical Edition. 1941;**13**:196-203. DOI: 10.1021/i560091a022

[52] Bryant WMD. Optical crystallographic studies with the polarizing microscope. IV. Axial dispersion with change of sign. Other dispersion measurements. Journal of the American Chemical Society. 1943;**65**:96-102. DOI: 10.1021/ja01241a031

[53] Polavarapu PL. Chiroptical Spectroscopy: Fundamentals and Applications. 1st ed. Boca Raton: CRC Press; 2017. p. 448. DOI: 10.1201/9781315374888

[54] Srebro-Hooper M, Autschbach J. Calculating natural optical activity of molecules from first principles. Annual Review of Physical Chemistry. 2017;**68**: 399-420. DOI: 10.1146/annurev-physchem-052516-044827

[55] Kaminsky W, Weckert E, Kutzke H, Glazer AM, Klapper H. Non-linear optical properties and absolute structure of metastable 4-methyl benzophenone.

Crystal Materials. 2006;**221**:294-299. DOI: 10.1524/zkri.2006.221.4.294

[56] Murphy VL, Kahr B. Hückel theory and optical activity. Journal of the American Chemical Society. 2015;**137**: 5177-5183. DOI: 10.1021/jacs.5b01763

[57] Murphy VL, Reyes A, Kahr B. Aromaticity and optical activity. Journal of the American Chemical Society. 2016; **138**:25-27. DOI: 10.1021/jacs.5b11138

[58] Molina P, Vasyliev V, Víllora EG, Shimamura K. CeF_3 and PrF_3 as UV-visible faraday rotators. Optical Express. 2011;**19**:11786-11791. DOI: 10.1364/OE.19.011786

[59] Vasyliev V, Víllora EG, Nakamura M, Sugahara Y, Shimamura K. UV-visible faraday rotators based on rare-earth fluoride single crystals: $LiREF_4$ (RE = Tb, Dy, Ho, Er and Yb), PrF_3 and CeF_3. Optical Express. 2012;**20**:14460-14470. DOI: 10.1364/OE.20.014460

[60] Víllora EG, Shimamura K, Plaza GR. Ultraviolet-visible optical isolators based on CeF_3 faraday rotator. Journal of Applied Physics. 2015;**117**:233101-233104. DOI: 10.1063/1.4922497

[61] Shinagawa K. Faraday and Kerr effects in ferromagnets. In: Sugano S, Kojima N, editors. Magneto-optics. 1st ed. Berlin: Springer; 2000. pp. 137-177

[62] Takanabe A, Koshima H, Asahi T. Fast-type high-accuracy universal polarimeter using charge-coupled device spectrometer. AIP Advances. 2017;**7**: 025209-025214. DOI: 10.1063/1.4977440

[63] Tanaka M, Yagi-Watanabe K, Kaneko F, Nakagawa K. First observation of natural circular dichroism spectra in the extreme ultraviolet region using a polarizing undulator-based optical system and its polarization characteristics. Journal of Synchrotron Radiation. 2009;**16**:455-462. DOI: 10.1107/S0909049509012291

Chapter 9

Third-Order Nonlinearity Measurement Techniques

Allen Moses Samuel Elizabeth

Abstract

To measure the degenerate (single-frequency) optical nonlinearities, third-order nonlinearity measurement and their related techniques were employed. When a laser beam is induced on a nonlinear (NL) medium, a phase change is easily identified using third-order nonlinearity measurement techniques (Z-scan). When the sample material is scanned on Z-axis, the phase change is denoted by sign and magnitude, the refractive index which is directly related to the change in the index of refraction. The nonlinear absorption from the absorption coefficient is independent of the index of refraction which is a required parameter for calculating nonlinear refraction. Further, the change in transmission caused by nonlinear absorption of the subjected material is related to the change in absorption coefficient which is easily determined by the Z-scan technique. From Z-scan responses, real and imaginary parts of third-order nonlinear susceptibility (χ^3) can be determined. The Z-scan technique is an interesting process that leads to optical power limiting and nonlinear optical propagation.

Keywords: nonlinear optics, nonlinear absorption, nonlinear refractive index, nonlinear absorption coefficient, and nonlinear susceptibility

1. Introduction

To examine the irritation dynamics and time-resolved technique under sonorant agitation, Z-scan is an effective tool. The intensity-dependent nonlinear (NL) absorption coefficient ($\Delta\alpha$) and magnitude of the nonlinear index (n_2) (Kerr nonlinearity) are vital key factors for photo-induced dissociation in reduced crystals, optical material, and photonic applications [1, 2]. Nonlinear absorption (NLA) and nonlinear refraction (NLR) in solid and liquid can be effectively measured using the Z-scan technique [3]. This method is relatively simple due to its single-beam technique measuring both the sign and magnitude of NLR and NLA, and its spatial beam distortions principle is purely dependent on the sign of the nonlinearity [2]. Methods like beam distortion [4], elliptical polarization [5], multi-wave mixing [6], nonlinear interferometry [7], the indexing change (Δn), and absorption change ($\Delta\alpha$) can be done directly without curve fitting. Z-scan method is the most sensitive mechanism that directly determines the refractive index change. The electronic structure of materials is shed by the frequency dispersion study from the nonlinear susceptibility (χ) due to the internal atomic and molecular resonances. The advantage reveals that

Z-scan studies become a fundamental screening for third-order nonlinearity [3, 8]. The sample is placed on the Z-axis along with the positive and negative directions, and the input laser beam was converted into polarized Gaussian beam along the waist of the motorized translation stage [9] as shown in **Figure 1**.

2. Technique

The usual "closed aperture" (CA) Z-scan technique (i.e. aperture in place in the far field) for nonlinear refraction is shown in **Figure 1** (sample displaying self-focusing). Sample transmittance of polarized Gaussian beam through the aperture is monitored by far field as a function of the Z position of NL in the vicinity of the linear optics focal position [10]. The scan range for the examination of the sample depends on the beam parameter and thickness of the sample L. A significant limitation is the diffraction length of the focused beam defined as a polarized Gaussian beam. A "thin" sample has a thickness of L. Even though all the information is theoretically contained within a scan range of $\pm Z$ direction, it is preferable to scan the sample for approximately ± 5 Z to determine the linear transmittance [2]. If the subjected material surface is rough or imperfect, it leads to background noise in the output. A reference detector can be used to monitor the normalized transmittance. To eliminate the possible noise due to spatial beam variations, this reference arm can be further to include a lens and an aperture identical to those in the nonlinear arm [11] as shown in **Figure 1**.

3. Interpretation

A typical CA Z-scan output for a thin sample exhibiting nonlinear refraction peak (T_P)–the valley (T_V) is shown in **Figure 2**, where $\Delta T_{P\text{-}V}$ is the difference between transmittance peak and valley transmittance. A self-defocusing nonlinearity results in a peak followed by a valley in the normalized transmittance as the sample is moved away from the lens (i.e. increase in the Z direction).

The normalization is performed in such a way that the transmittance is unity for the sample far from the "focus where the nonlinearity is negligible [12]. The negative lensing in the sample placed before the focus moves the focal position further from

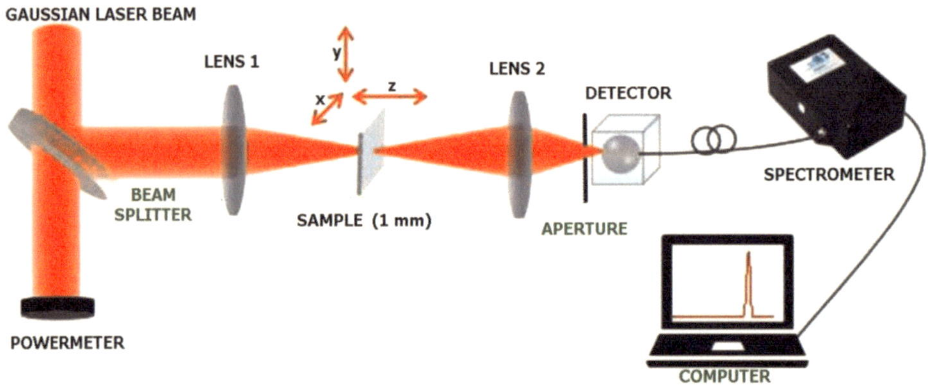

Figure 1.
Experimental arrangement of Z-scan technique.

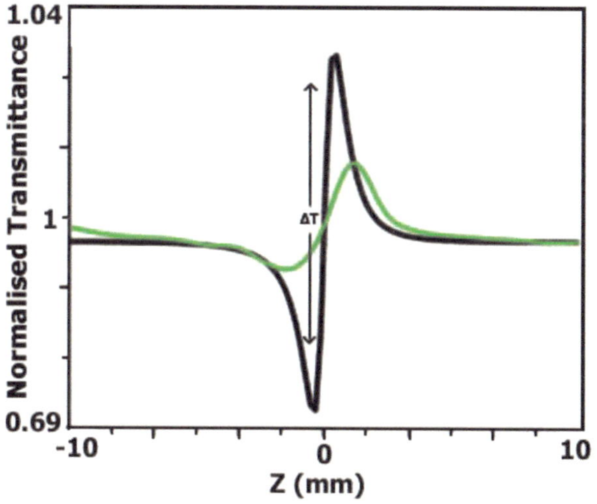

Figure 2.
Z-scan for third-order positive nonlinearity.

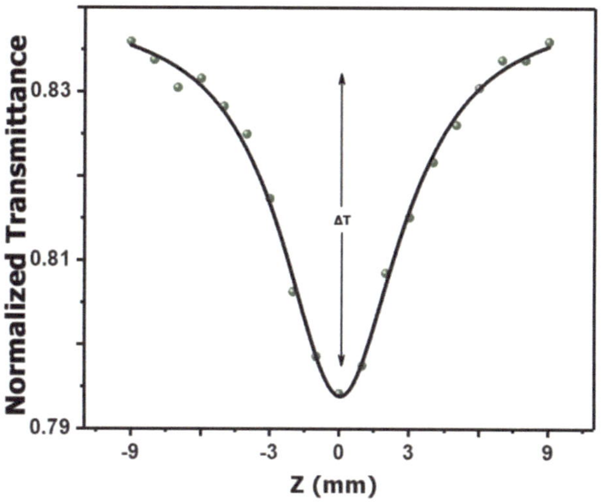

Figure 3.
Z-scan graph of a thin nonlinear absorber.

the sample placed increasing the aperture transmittance. The experiment with T = 1 is referred to as an "open aperture" (OA) Z-scan and allows direct measurement of nonlinear absorption in the sample as shown in **Figure 3**.

4. Nonlinear refraction

Nonlinear refraction (NLR) in the absence of nonlinear absorption (NLA) is determined by using a finite aperture to calculate the transmittance of the NL medium in a far field as a function of the position (Z) of the sample. The illumination of the sample is done by the self-focused polarized Gaussian beam that attains different incident field strengths at different positions of Z [3]. Obtained values can be convenient to plot T, the transmittance normalized to the linear transmittance of the system. The nonlinearity can be estimated from the difference between the maximum (peak) and minimum (valley) values of the normalized transmittance (ΔT). For a thin optical Kerr medium where the refractive index varies linearly with irradiance nonlinear refractive index coefficient n_I, ΔT is proportional to the nonlinear phase distortion (shift) on the axis with the sample at the waist [13, 14]. The empirically determined relation between the induced phase distortion ($\Delta\phi_0$) and normalized transmittance ($\Delta T_{P\text{-}V}$) for a third-order nonlinear refractive process in the absence of NLA is given by Eq. (1):

$$\Delta T_{p-v} \cong 0.4061(1-S)^{0.27}|\Delta\varphi| \tag{1}$$

where S is the linear transmittance of the aperture in the absence of the sample and $\Delta\phi$ is the axis phase shift.

$$S = 1 - \exp\left[\frac{-2r_a^2}{\omega_a^2}\right] \tag{2}$$

Here, r_a is the radius of the aperture and w_a is the beam radius at the aperture.

$$\Delta\varphi_0 = kn_2I_0L_{eff} \tag{3}$$

Here, L_{eff} is an effective sample length and $k = 2\pi/\lambda$, so,

$$\Delta\varphi_0 = \frac{2\pi}{\lambda}n_2I_0L_{eff} \tag{4}$$

Here, λ is the wavelength, n_2 is the nonlinear index of refraction, and I_0 is the axial irradiance at the waist.

$$L_{eff} = \frac{1-e^{-\alpha L}}{\alpha} \tag{5}$$

where L is the sample length, $L = L_{eff}$ (absence of linear absorption), and α is the linear absorption coefficient. The distance between peak and valley is measured in Z-axis, and $\Delta T_{P\text{-}V}$ is a direct measure of the diffraction length of the incident beam for a given order nonlinear response. In a standard Z-scan (i.e. using a Gaussian laser beam and a far-field aperture), Eq. (6) gives the relation for third-order nonlinearity (Z):

$$|\Delta Z_{pv}| \approx 1.7Z_0 \tag{6}$$

This gives the focal spot size of the beam for diffraction-limited optics independent of the irradiance for small nonlinearities.

In principle, the Z-scan can be used to measure very small spot sizes using a very thin sample. Nonlinear absorption measurements are usually done by removing the aperture to collect the maximum intensity from the sample.

5. Higher-order nonlinearities

The nonlinear optical effects give index changes proportional to the irradiance (Δn). For a fifth-order ($\chi^{(5)}$), NLR becomes the dominant mechanism in semiconductors when Δn is induced by two-photon generated free carriers. This type of nonlinearity is derived from simple relations that accurately characterize the Z-scan data [15]. A Gaussian beam and far-field aperture is given by Eq. (7):

$$\Delta T_{pv} \cong 0.21(1-S)^{0.27}|\Delta\varphi_0| \tag{7}$$

and

$$|\Delta Z_{pv}| \approx 1.2Z \tag{8}$$

For this case, the data analysis becomes very complicated due to the simultaneous process of $\chi^{(3)}$and $\chi^{(5)}$ using several Z-scans at different irradiances [15]. This procedure makes use of simple relations of Eqs. (1) and (3) to estimate the nonlinear coefficients associated with both $\chi^{(3)}$and $\chi^{(5)}$ processes.

6. Nonlinear absorption

The NLA can be determined using a two-parameter fit to a closed aperture Z-scan, and it is more accurate to determine in an open aperture of Z -scan. For small third-order nonlinear losses with response times much less than the pulse width (i.e. two-photon absorption), and for a Gaussian temporal shape pulse, the normalized change in transmitted energy ΔT with Z is given by Eq. (9) [16]:

$$\Delta T \approx -\frac{q_0}{2\sqrt{2}}\frac{1}{[1+Z^2/Z_0^2]} \tag{9}$$

The Lorentzian distribution of the illuminance with Z position for a focused polarized Gaussian beam is shown in **Figure 3**. If the response time is extended than the pulse width, the factor $2\sqrt{2}$ is replaced by 2, which is independent of the temporal pulse shape. This gives the excited state of absorption cross section (σ).

7. Thin nonlinear medium

For an absorption coefficient that varies linearly with irradiance, the coefficient of nonlinear absorption α can be calculated using Eq. (11):

$$\alpha = \frac{q_0}{I_0 L_{eff}} \tag{10}$$

$$\Delta T = \left|1 - \frac{1}{q_0} 1n(1+q_0)\right| \tag{11}$$

Here, I_0 is the incident beam intensity, and Eq. (3) can be used for all orders of I_0. The Z-scan experiment is a simple method due to its single-beam technique. The Z-scan technique yields both the sign and magnitude of the nonlinearity from n_I and α. In addition, it has an advantage process of a closer similarity between Z-scan and optical power limiter geometries [17]. A comprehensive Z-scan study not only gives important information on the NL properties of the sample but also yields necessary information regarding optimization of the optical power limiting geometry such as the optimum sample position and optimum sample thickness. This makes Z-scan an ideal tool for assessing materials for optical power-limiting applications.

8. Modeling studies

The Z-scan technique is a far-field measurement method. The term "far field" is defined as a distance of 10 Rayleigh lengths between the beam waist and the aperture, and there is an 18% change in the transmittance. An eliminating error in ΔT is only 1% because both the peak and valley are calculated in the Z direction [3].

The poor aperture alignment was effectively determined by modeling studies for the subjected sample Z-scan profile. For best results, the sample should be wedged and alignment of the aperture focused on words of the center of the sample to obtain a maximum transmittance through the sample. Modeling studies also launch the effects which ensue if the sample has a lens-like profile or if it shows surface roughness or scratches. Implementing the technique of subtracting low-power from high-power Z-scans, as suggested by Sheik-Bahae et al. [2] works well for extracting the NL result for imperfect surfaces.

The refracted beam from the sample gives rise to the third-order nonlinearity, and it is intensity-dependent convergence or divergence (self-focusing/defocusing). Sample position Z causes intensity variation due to the transformation of phase distortion converted to amplitude distortion of the transmitted beam [3]. As the position of the Z varies, if there is an increase in transmittance in the pre-focal stage and followed by a decrease in the post-focal region (peak-valley) Z-scan symbolizes negative NLR, whereas a valley-peak represents a positive nonlinearity. Removing the aperture leads to collecting total intensity on the detector which results in the Z-scan for flat response (purely refractive nonlinearity) due to the sensitivity of NLR depending on the aperture. The multi-photon absorption leads to valley enhancement quashes the peak saturation of absorption provides a reverse saturable absorption (RSA) effect in CA of Z-scan [18]. Significantly, this technique enables the measurement of the sign of nonlinearity by directly eliminating magnitude (real and imaginary parts). For optical signal processing applications, the sign of nonlinearity plays a key factor. It cannot be directly derived using any other techniques [19].

9. Z-scan measurement technique

Third-order NL parameters of the subject material were determined by the Z-scan technique. A wavelength of 532 nm (Gaussian laser beam) was converged using a converging lens of focal length 10 cm in the Z position of the subject sample with a thickness of 1 mm. The NL parameters such as absorption coefficient (β), refractive index (n), and susceptibility ($\chi^{(3)}$) were calculated using their equations. The subjected sample was displaced in the Z direction to obtain intensity-dependent absorption. The subjected sample information was recorded using a high-sensitive photo-detector [6]. A sample is drawn in between the obtained values, and a curve fit was done as shown in **Figure 4**.

The nonlinear absorption coefficient (β) was calculated using the following equation:

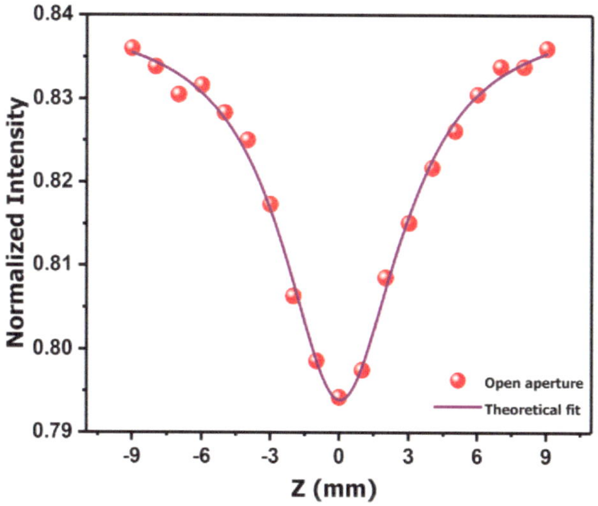

Figure 4.
Open aperture (OA) curve.

$$\beta = \frac{2\sqrt{2\Delta T}}{I_0 L_{eff}} (m/W) \tag{12}$$

Keeping the aperture in front of the detector Z-scan was performed to get the sign and magnitude closed aperture. The NLA and NLR were essentially important in closed aperture transmittance [14, 20–22]. The ratio between CA and OA gives the real NLR. A sample plot of the closed aperture (CA) curve is depicted in **Figure 5**.

To calculate the third-order nonlinear refractive index (n), Eq. (13) was used.

$$n = \frac{\Delta\varphi}{kI_0 L_{eff}} \left(m^2/W\right) \tag{13}$$

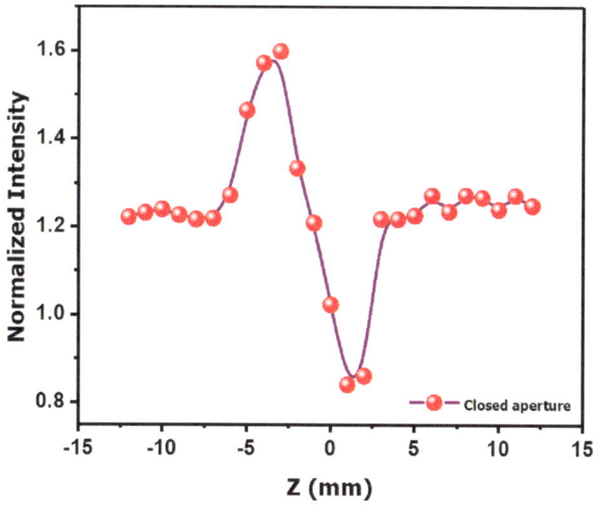

Figure 5.
Closed aperture (CA) curve.

The on-axis phase shift at the focus was indicated by $\Delta\varphi$; it has been found using the Eq. (14):

$$|\Delta\varphi| = \frac{\Delta T_{p-v}}{0.406(1-S)^{0.27}} \quad (14)$$

The obtained values of n_2 and β were used to calculate the complex equations containing the real and imaginary elements of third-order nonlinear optical susceptibility ($\chi^{(3)}$), which can be calculated using the Eqs. (15) and (16) [21]:

$$R_e\chi^{(3)}_{(esu)} = \frac{10^{-4}\varepsilon_0 C^2 n_0^2 n_2}{\pi}(cm^2/W) \quad (15)$$

$$I_e\chi^{(3)}_{(esu)} = \frac{10^{-2}\varepsilon_0 C^2 n_0^2 \lambda\beta}{\pi}(cm/W) \quad (16)$$

where ε_0 indicates the permeability of free space, C is the velocity of light, and the n_0 is the linear refractive index of the sample. The magnitude of third-order nonlinear susceptibility ($\chi^{(3)}$) for the subjected sample was estimated using the Eq. (17):

$$\chi^{(3)} = \sqrt{(R_e\chi^{(3)})^2 + (I_m\chi^{(3)})^2} \quad (17)$$

The NL of the subjected material purely depends on the concentration of the solvent because its intermolecular interaction Eq. (17) provides a better nonlinearity.

10. Conclusion

Several methods and techniques are available to determine the third-order nonlinearity measurement techniques; among them, Z-scan is the simplest and most efficient technique. The Z-scan technique directly measures the physical processes behind the nonlinear response of a given material using a single wavelength. Nonlinear absorption and refraction invariably coexist because they are obtained from the same physical parameters. They are related by dispersion relations identical to the usual Kramers-Kronig relation that connects linear absorption to linear index, or equivalently, leads to real and imaginary parts of linear susceptibility. And also it has both a completely computed technique for determining standards and a relative measurement method. The Z-scan signal is a function of irradiance and shapes for sample position. The Z-scan technique has a great prospect to solve highly scattering problems and surface the way to characterize the NLO properties of biological and optical polished samples. It can give useful information on the order of nonlinearity as well as its sign and magnitude.

Acknowledgements

I am very much grateful to my institution VINAYAKA MISSION'S RESEARCH FOUNDATION, Salem, for providing an opportunity to carry out the research work in the Department of Physics, School of Arts and Science, Av campus, Payinoor,

Chennai. My heartfelt thanks to Late Dr. SM. Ravikumar, for supporting me throughout my research. I thank IIT Madras for the charazation analysis. I extend my gratitude to thank Dr. Samuel, Dr. Annie, Ms. Ivy, and Ms. Ezhil for their valuable suggestion.

Conflict of interest

The authors declare no conflict of interest.

Compliance with ethical standards

This study was approved by the School of Arts and Science, VINAYAKA MISSION RESEARCH FOUNDATION, AV campus, Chennai.

Author details

Allen Moses Samuel Elizabeth
Department of Physics, School of Arts and Science, Vinayaka Mission's Research Foundation, Av-Campus, Chennai, Tamilnadu, India

*Address all correspondence to: allenmoses.se@gmail.com

References

[1] Mills DL. Nonlinear Optics: Basic Concepts. Springer Science & Business Media. Berlin, Heidelberg, New York: Springer-Verlag; 2012

[2] Sheik-Bahae AAS, Van Stryland EW. High-sensitivity, single-beam n_2 measurements. Optics Letters. 1989;**14**: 955

[3] Sauter EG. Nonlinear Optics. New York: John Wiley & Sons, Inc.; 1996

[4] Williams WE, Soileau MJ, Van Stryland EW. Optical switching and n2 measurements in CS2. Optics Communication. 1984;**50**:256-260

[5] Owyoung A. Ellipse rotation studies in laser host materials. IEEE Journal of Quantum Electronics. 1973;**9**:1064-1069

[6] Smith PW, Tomlinson WJ, Eilenberger DJ, Maloney PJ. Measurement of electronic optical Kerr coefficients. Optics Letters. 1981;**6**: 581-583

[7] Milam D, Weber MJ. Measurement of nonlinear refractive-index coefficients using time-resolved interferometry: Application to optical materials for high-power neodymium lasers. Journal of Applied Physics. 1976;**47**:2497-2501

[8] Allen Moses SE, Tamilselvan S, Ravi Kumar SM, Vinitha G, Hegde TA, Vimalan M, et al. Synthesis, growth and physicochemical properties of new organic nonlinear optical crystal L-threoninium tartrate (LTT) for frequency conversion. Materials Science for Energy Technologies. 2019;**2**:565-574

[9] Allen Moses SE, Tamilselvan S, Ravikumar SM, Vinitha G, Hegde TA, Shanmuga Sundar GJ, et al. Crystal structure, spectroscopic, thermal, mechanical, linear optical, second-order and third-order nonlinear optical properties of semiorganic crystal: L-threoninium phosphate (LTP). Journal of Materials Science: Materials in Electronics. 2019;**30**:9003-9014

[10] Krauss TD, Wise FW. Femtosecond measurement of nonlinear absorption and refraction in CdS, ZnSe, ZnS. Applied Physics Letter. 1994;**65**:1739-1741

[11] Van Stryland EW, Sheik-Bahae M. Z-scan technique for nonlinear materials characterization. In: Proc. SPIE 10291, Materials Characterization and Optical Probe Techniques: A Critical Review, 102910Q. San Diego, CA, United States: Event: Optical Science, Engineering and Instrumentation; 1997

[12] Davy JR, Rees SR, Staromlynska J, Hermann JA, Gillyon MP, McKay TJ, et al. Optical devices utilizing nonlinear organic materials. In: Prasad PN, Mark JE, Fai TJ, editors. Polymers and Other Advanced Materials. Vol. 786. New York: Springer, Plenum Press; 1995. pp. 405-419

[13] Thakur MK, Quintero-Torres R. Sign and magnitudes of the off-resonant nonlinearities of polydiacetylene crystal measured by z-scan. In: Proc. SPIE 2025, Nonlinear Optical Properties of Organic Materials VI. San Diego, CA, United States; 1993

[14] Allen Moses SE, Johnson J, Nagaraju P, Raghu Y, Shanmugavalli R, Gunaseelan R, et al. Crystal engineering and physicochemical properties of l-cysteine cadmium chloride (LCC) for frequency-doubling and optical limiting applications. Journal of Materials Science: Materials in Electronics. 2022; **33**:1489-1502

[15] Said AA, Wamsley C, Hagan DJ, Van Stryland EW, Reinhardt BA, Roderer P, et al. Third- and fifth-order optical nonlinearities in organic materials. Chemical Physics Letters. 1994;**228**: 646-650

[16] Gang G, Zhang W, Zen H, Youwei D, Han Y, Zhang W, et al. Large non-linear absorption in C60 thin films. Journal of Physics B: Atomic, Molecular and Optical Physics. 1993;**26**:451-462

[17] Li L, Yuan HJ, Hu G, Palffy-muhoray P. Third-order optical non–linearities of nematic liquid crystals. Liquid Crystals. 2006;**16**:703-712

[18] Boyd RW. Nonlinear Optics. Rochester, New York: Academic Press; 2013

[19] Stegeman GI, Stegeman RA. Nonlinear Optics: Phenomena, Materials and Devices. Hoboken, New Jersey: John Wiley & Sons, Inc; 2012

[20] Kubendiran T, Ravi Kumar SM, Allen Moses SE, Nasareena Banu A, Shanthi C, Sivaraj S. Second and third-order nonlinear optical, mechanical, surface characteristics of Bis (thiourea) manganese chloride (BTMC) grown by slow cooling technique used for frequency conversion applications. Journal of Materials Science: Materials in Electronics. 2019;**30**:17559-17571

[21] Moses SEA, Kumar SMR, Hegde TA. Crystal engineering and physicochemical properties of L-threoninium succinate (LTS) single crystal for frequency conversion applications. Journal of Materials Science: Materials in Electronics. 2020;**31**:21097-21107

[22] Johnson J, Moses SEA, Srineevasan R, Kumar SMR. Materials synthesis, crystal growth and third nonlinear optical properties of semiorganic crystal: 4-dimethylaminopyridine lithium nitrate (DMAPLN) crystal. Journal of Materials Science: Materials in Electronics. 2021;**32**:22187-22195